A Practical Guide to
TECHNICAL REPORTS
AND PRESENTATIONS
for Scientists, Engineers, and Students

Pauline Bary-Khan

Elizabeth Hildinger

Erik Hildinger

Custom Publishing

New York   Boston   San Francisco
London   Toronto   Sydney   Tokyo   Singapore   Madrid
Mexico City   Munich   Paris   Cape Town   Hong Kong   Montreal

Cover Art: Courtesy of PhotoDisc/Getty Images

Printed in the United States of America

3 4 5 6 7 8 9 10   V3CR   13 12 11 10 09

2008780024

LM

Pearson
Custom Publishing
is a division of

www.pearsonhighered.com

ISBN 10: 0-555-01787-7
ISBN 13: 978-0-555-01787-6

CONTENTS

APPENDICES

INTRODUCTION

This book is meant to help technical and scientific professionals with technical writing. Also, it is meant to be handy and easy to use. It focuses on practice rather than on theory. However, its text, advice and examples should be sufficient for most projects. Readers will find plain advice and unqualified statements about how to write; for the sake of brevity, we will devote very little space to justifying the principles and advice in this book. In any case, the reasons for the advice should be evident if you read the text thoroughly.

Technical writing treats technical, industrial or scientific matters; the writing must therefore be precise, clear, and concise. Examples of technical writing are memoranda, proposals, laboratory reports (test reports), progress reports, design reports and research reports.

We want to give you a tool that you can use to succeed in the communication tasks— primarily writing reports and giving oral presentations—that you will encounter in your professional career.

Organization of the Book

This book is organized to be simple and efficient to use. It falls into four distinct sections:

General Principles of Technical Communication
Following the introduction, the first chapter presents general principles of technical report writing and recommendations for structure and language that are relevant and applicable to many technical reports.

The Basic Memorandum
The second chapter presents the memorandum, because mastery of that form will teach you a number of concepts that apply to other forms of

technical writing. The chapters that follow treat proposals, laboratory and test reports, progress reports, design reports, research reports, and oral reports. Your work may require you to produce a report of a type not covered specifically in this book; however, if you examine the purpose and organization of the reports that have been included, you should find one that, with a little modification, will work for you.

Other Types of Reports

Following the chapter on the memorandum are chapters on the other types of reports; each type is described and then followed by a template and examples that correspond to the descriptive text. This arrangement should allow you to grasp the structure easily and then apply it to the task you face.

Appendices

You will find appendices at the end of this book that treat a number of important matters and that will lead you to further, more detailed texts, if you wish to pursue them in more depth. These appendices treat

1) organization, development of content, and argumentation;
2) strategies for developing defensible claims, selecting good evidence, and making effective use of source material;
3) strategies for achieving a clear and concise style;
4) rules and guidelines for grammar, sentence structure, diction, and punctuation;
5) strategies for development of effective graphics;
6) guidelines for professional use of e-mail;
7) features of formal reports.

The appendices provide editing and revising checklists as well. We urge you to read the appendices because they include additional information that you can use to fine-tune your report writing skills.

1

GENERAL PRINCIPLES OF TECHNICAL COMMUNICATION

This chapter contains general principles for technical reports, both oral and written. Much of the material is most directly applicable to written reports; for principles specific to oral presentations, see Chapter 8.

1.1 AUDIENCE AND PURPOSE

You need to recognize two crucial ideas when you are creating a technical document or presentation. Considering these broad ideas will help you make decisions about handling a number of aspects of any writing job, such as diction, tone, level, backgrounding, and organization.

Audience

The first crucial concept is *audience*—that is, the reader or readers of whatever you write. You must first always think about who it is that you are writing for and, once you have determined who your audience is, think what that implies. When you know who the reader of your document is, ask yourself a series of questions. Here are some reasonable ones to start with:

- What does this reader know about the subject or situation already?
- Is the reader inclined to agree or disagree with me about what I will write?
- Does the reader have a stake in what I'm writing about?
- What actions does the reader need to take on the basis of my document?

We cannot possibly list all the relevant questions; they will depend on the situation and the task you have before you. But you must find the right questions and then answer them. The answers to these questions will suggest to you how you should organize the writing, what background information you should provide, what tone you should take, and what sort of arguments you may need to develop and advance.

Mixed Audiences

As if the previous material weren't complicated enough, there's another layer: you must consider the possibility of a *mixed audience*—one made up of people with different experiences, different levels of education, different interests in a given project and different cultural backgrounds. In professional settings, the documents you write are rarely seen by only one person; more often, they'll go to more than one reader and be read for more than one purpose.

Purpose

That brings us to the second crucial concept that you must keep in mind: *purpose*—in other words, the reason you are writing your document. Are you responding (as you frequently will be) to a request for information, solutions, or recommendations? Is your report intended primarily to persuade or to inform? If it's a persuasive report, are you trying to get someone to give you funding for something or permission to do something? Are you trying to persuade someone to believe you, trust your recommendations, or put a project into your hands? If it's informative, are you reporting to someone or to a committee about something you have learned? Sometimes your purposes are mixed: you may be providing information and attempting to bring about some action on the basis of or in response to that information. Or sometimes you may be telling your manager something he or she won't want to hear. These last two examples show particularly clearly how audience and purpose are interrelated; these interrelationships will lead you to make certain decisions about tone and organization, as well as content.

1.2 ORGANIZATION OF TECHNICAL REPORTS

Now that we have discussed audience and purpose, let's turn to the organization of technical reports. The fundamental principle is this: *information, to be understood, must be organized in some purposeful manner.* Without a pattern of organization, the writing may become difficult to understand or laborious to read.

First, the information must be framed in such a way that the reader can grasp it. In other words, if you are communicating new information (information unknown to the readers) you cannot simply introduce it to your readers and expect them to understand it. Instead, you must begin your discussion with information the reader already knows and then connect it to the new information. Therefore, providing adequate background is critical.

Second, it is generally useful to organize information (in default of any other structure) from the general to the specific. Begin with a broad statement and then give details and qualifications. This principle applies not only to the construction of paragraphs but also to the organization of whole sections in a report. In other words, go from broad to narrow: explain general concepts first and then move to more detailed information. Paragraphs or sections of reports that follow this rule will be easier to understand as a result.

When you have reached a certain level of specificity, as you usually will in technical documents of any length, you may find that the rule above becomes difficult to apply. In that situation, you will need to use one of several other forms of organization that are suited to particular purposes. We will talk further about these in the next section.

The Use of Headings and Sub-headings

In technical reports, the high-level organization is usually signaled by headings and subheadings (as you will see in the sample reports in the following chapters). These should be informative, and the headings within sections (if not throughout) should be grammatically parallel. That is, if your first heading is a noun phrase, such as, for example, 'Components of the X-1200-G Widget,' your next heading should be in the same form: for example, 'Functions of the X-1200-G Widget,' rather than, say, 'How the X-1200-G Widget Works' or 'To Operate the X-1200-G Widget.' While this may seem to be a grammatical point of little importance, in fact parallelism is an important signal to your reader. It is important to use this means of conveying structural information.

1.3 ELEMENTS OF TECHNICAL REPORTS

Technical reports of many different sorts tend to be quite similar in most of their constituent elements. The similarity is the result of their shared primary purpose: to communicate in a useful and efficient way to a mixed audience whose reasons for reading the documents may vary but will almost invariably be pragmatic. The main elements of most technical reports are the overview, the report body, and the appendices. We'll talk briefly about each one below.

Overview/Front Matter

At the beginning of most technical documents, you will find a section of material that goes by several different names and takes slightly different forms but has one important purpose: to acquaint the readers with the occasion for the document—that is, what problem did the writer(s) address? Why was the problem addressed? Why is it important?—and let them know what the document's contents are. Thus this section, whether it is just called 'overview' or 'front matter' or goes by another name, such as 'summary,' 'executive summary,' 'foreword and summary,' or 'abstract,' will almost invariably contain the same or similar elements.

Try not to be confused by the variety of names; just focus on the elements:

First element: Problem statement
A statement of the institutional or organizational problem
A statement of the importance of the problem
A statement about the assignment the writer or writers were given

Second element: Task statement
A statement of what the writer(s) did in response to the problem
A statement of how the writer(s) performed their task(s)

Third element: Results, conclusions, findings, recommendations, and implications for the reader
A statement of what the writer(s) found in the course of carrying out the task(s)
A statement of the evidence for any conclusions
A statement indicating the basis of any recommendations
An enumeration of any implications for the managerial reader (costs, scheduling or personnel requirements, etc.)

Fourth element: Document function statement
A statement indicating what the document is for and how it is to be used
Sometimes, an indication of the content and organization of the document

As you can see, the elements consist of the information necessary to orient a managerial reader to the problem you, the writer, addressed and to prepare him or her for what will follow. An overview containing this set of information enables a busy reader to tell very quickly what problem the document addresses and whether and how the problem was solved; it also gives him or her a heads-up as to any remaining concerns or complications that will ensue.

Your institution or organization will probably tell you what type of overview section you should use for your technical reports. We've listed them below in the order of their likely usefulness to you:

Executive summaries, which present all of the elements, the problem, the task, the results, conclusions and/or recommendations, and the document function, usually in the order they appear above, are typical of many types of engineering documents in industry—see the sample design report in Chapter 6. They may be broken into several paragraphs or contain bulleted lists.

Foreword and summary overviews, which contain the elements above but break them into two sections and reorder them, are also typical of many engineering documents in industry. The *foreword* section contains the first two elements, the problem and the task, and the last element, the document function, usually in a very condensed form. Its specific purpose is to orient the reader to the problem and prepare him or her for what will follow. The *summary* section contains (or consists of) the third element, the results, conclusions and/or recommendations. Its specific purpose is to present the reader with the crucial content of the following report in a highly condensed form—see the sample memorandum in Chapter 2.

Abstracts, which may not contain full forms of all of these elements and usually focus on the problem statement, the task statement, and a forecast of the organization, are typical of research papers such as the sort you will find in engineering journals. They are sometimes found as the overview section of laboratory reports as well. Abstracts typically take one of two forms: the *descriptive abstract*, which states the problem and tells what the report contains but does not present any conclusions, or the *informative abstract*, which states the problem and presents the main conclusions or findings along with a forecast of the report's contents. Institutions and journals usually favor one or the other; to find out which you should use, you'll need to look at examples of documents in your company's files or in a recent issue of the journal you're submitting your research paper to.

Stand-alone summaries, which typically lack some of the material specifically directed to a managerial audience, are not common in professional engineering documents, but they may occur at the beginning of short reports or long memoranda. They usually contain just a brief statement of purpose, a condensed version of the contents, and a forecast of the organization.

Further Audience Considerations

Three important and specific audience considerations influence the way you should write your overviews.

First, you can't assume that your audience is fully informed as to the nature of the problem and your role with regard to it. Even if your primary audience does know the background and current situation (and a busy manager with many people working under him or her may need a quick memory jog even so), remember that you are usually writing for a secondary audience or audiences who may be unfamiliar with the circumstances that occasioned your report as well. These readers will have difficulty making use of your document if it doesn't provide clear problem and task statements.

Second, you can't assume that your audience always shares your technical expertise. In fact, you should assume that your audience does **not** know as much as you do about the subject you are writing about. Therefore, you must use language that a non-expert can understand, and if highly technical terms are unavoidable you must define or explain them.

Third, you can't assume that your audience will read the entire document thoroughly. Many managers read the overview and the skim the remainder, or hand it off to someone else to read in detail. Therefore, you must include information that the reader needs in order to make decisions or to understand the problem in your overview. A comprehensive summary is essential: a summary that omits crucial information may cause major problems for your reader.

The Body of the Technical Report

The nature and purpose of your technical report will determine the specifics of its content, but some elements will be consistently present: you will almost certainly need an introductory section, the section(s) presenting the technical content, and a concluding section. We will talk briefly about each of these sections below.

Introductory Sections

Even though you have inserted an overview into your technical document, you still need an introduction. Overviews have a different purpose and a different intended audience; the reader who reads your overview may not even read the body of the report and vice-versa. If a reader received your report without its overview, what information would he or she need to find to get started reading with understanding? Answering that question will lead you to a description of the introduction.

Although the content will vary somewhat depending on the type of report, almost all introductions will contain the following (you'll see some overlap between the overview and the introductory section, but that's perfectly appropriate):

Elements of an introductory section
- Context
- Problem and its importance (including current research)
- Scope: the specific topics examined in the report
- Forecast: contents and organization of document

Context is the term we use for the information that helps a reader 'locate' the document and its contents in a field of research, a stream of communications, or another setting. For example, a research or design report will probably begin by introducing the broad area to which the research or design belongs. Its purpose is to make the statement of the problem comprehensible to the reader. The other elements should be self-evident; for examples, see the sample reports in **Chapters 6 and 7**.

Content Development and Argumentation

No matter what type of technical report you are writing or what its subject may be, you know that it has to contain adequate content. You may assume that this content will be technical material, and that's right, to an extent. But consider: you can't just put a mass of technical details down on a piece of paper, or present them to an audience, without imposing order on them. Furthermore, technical details alone can't do very many of the jobs your technical documents need to do; by themselves, they can rarely explain anything adequately, and they can almost never persuade a reader to do anything. Technical details need to work as part of the support or evidence for broader statements, and these broader statements have to be shaped and crafted to suit the larger purposes of the report. Thus you need to think about **content development**, and it's most helpful to think of that as a combination of four interrelated processes:

- Selecting the best structure (mode of development)
- Developing defensible claims
- Selecting good evidence
- Making effective use of source material

Combined, these constitute the process we'll call **argumentation**. Argumentation (as distinct from *persuasion*) is the main mode of academic and professional discourse. In brief, here's what these processes involve; for more detailed discussion, see **Appendix A**.

Selecting the Best Structure: Sections and paragraphs within a report present their contents in a number of fairly well-defined patterns, according to the purpose they need to serve: classification of a set of items, causal analysis, comparison/contrast, description of an object or process, or narration of a series of events, to name a few. Once you have determined what the purpose of any given section of your technical report is, you need to construct a *topic sentence* that will control or direct the remaining contents. You also need to supply connective words and phrases that indicate the development you are using; for example, in a comparison/contrast section you'll probably begin several sentences with phrases like 'in comparison,' 'similarly,' and 'in contrast.' Any good college handbook can expand your repertoire and sharpen your use of such expressions.

Developing Defensible Claims: In technical reports, claims most often represent a conclusion that the writer has reached through investigation, research, or reasoning and that he or she wants the reader to accept as valid. The assumption is that you will support your claims (we'll discuss that below); thus you must take care to shape your claims so that you can defend them by means of the kinds of support that engineers and other scientifically-minded people will accept. The guidelines, then, that you need to follow in crafting your claims are straightforward:

- Claims must have verifiability and falsifiability. This guideline excludes tautologies, mere statements of personal taste, and observations of physical fact.
- Claims must be as concrete and as specific as you can make them. This guideline excludes universal statements and statements so general that they are indefensible in the space you have available.

For more details on the nature of claims, see **Appendix A.** Any good textbook on logic, argumentation, and rhetoric will also give you a great deal of useful information on this subject.

Selecting Good Evidence: The best evidence in most engineering documents is quantitative; most conclusions you reach will rest on measurements, calculations, experimentation (yours or that of researchers whose results you find), and the like. If you have numerical data to support your claim, you should present it—as clearly, thoroughly, and accurately as you can.

Not every claim, of course, arises from operations with numerical data. If you are establishing criteria for success of a design, for example, you may have a different sort of data, perhaps reports of how comparable products have fared in the marketplace or statements made by the client in the call for proposals. If you are claiming that a particular material is suitable for use in a particular application, you may support it (at least in part) with a description of the environment in which that application will be used. If you are establishing causality, you may use straightforward logical deduction to eliminate possible causes until you arrive at one. In any of these and other cases, two things remain constant:

- Evidence, or the basis for your belief that the claims are reliable, is an essential element of your arguments;
- A connection between your evidence and the claim it supports must be clearly apparent to your audience.

For an explanation of this second point, see **Appendix A**. You can find fuller discussion of this and many related issues in any good textbook on argumentation; in particular, you can find useful material under the heading 'Toulmin model.' We encourage you to pursue the topic on your own; effective argumentation is one of the most valuable skills an engineering student or professional can acquire.

Making Effective Use of Source Material: Often the evidence or support for claims in technical documents comes from material the writer has located in the course of doing research on the topic. Effective use of such material depends on several things: the reader has to be able to tell what material comes from the source and what is original; the reader has to be able to make sense of the material, and the material has to be connected to the argument the writer is making. What this means in practice is that you need to know several things:

- how to credit sources
- how to use the graphic indicators (quotation marks, ellipsis, indentation, spacing, and parenthetical documentation) that convey information about what's being taken from sources

- how to paraphrase source material accurately
- how to incorporate quoted or paraphrased material

You also need to know one very important precept: no one else can make your argument for you—you have to construct the argument and then, where it's appropriate, insert material you've gathered through research to support the claims you've formulated. You'll find fuller explanations and guidelines to help you with the processes we've listed above in **Appendix A**.

Concluding Sections

Most technical documents will need a concluding section of some sort. These vary according to the purpose of the technical report. Some typical contents follow:

- Reiteration of main findings
- Implications of findings
- Actions to be taken by writers
- Actions suggested to reader(s)

Not all technical reports have formal conclusions; an informative report, for example, may have no implications or suggested actions, and its main findings will have been enumerated at the beginning. For examples of some different kinds of concluding sections, see the sample reports in Chapters 6 and 7.

References

The final body section of any technical report that incorporates source material is a list of references keyed to the citations within the text itself. The style you should use for this list depends on the nature of the document (and the preferences of your supervisor or instructor). Generally you should use one of the standards for the engineering profession: APA, ASME, or AMA style. Neither MLA nor Chicago style is common for engineering documents.

Full and accurate bibliographical information is essential for many reasons, not the least of which is the legal and ethical requirement to give credit to other authors when you use their work. For some examples and guidelines for incorporating source material see **Appendix A.**

Appendices of Technical Reports

Often the argumentation in a technical report relies on detailed data that would overwhelm the reader if it were included in the body. Such material—test results, e.g., or detailed specifications for alternative designs—is typically placed in a separate section called an appendix (or a series of appendices), which is the last item in your report and may even constitute a separate volume in very long reports.

The general rules you need to guide you are very simple:

- Put only non-essential matter into the appendices
- Be sure to refer your reader to the relevant appendix at the point that it is relevant
- Title it (or them) clearly
- Indicate the presence and content of your appendices in your memo heading or table of contents.

1.4 LANGUAGE AND STYLE IN TECHNICAL REPORTS

The most important feature of an effective technical report is usability, and if a document is going to be useable, the primary goal of the writer has to be readability. Readability is a product of several things, but you achieve it primarily through your organization and your language. We've talked about organization above; now we'll look at language and style. The focus of this section is a general examination of the choices you can make to achieve clarity and precision, concreteness, politeness of tone, and a suitable level of formality. **Appendix B** contains further details pertinent to several of these.

Grammar and Syntax

Grammar and syntax are important, because using Standard English helps you to establish credibility. Conversely, poor grammar and sentence structure can cause your reader to doubt your competence in other areas. These considerations alone should encourage you to use the language correctly.

The second reason for the importance of grammar and syntax is that English has a relatively complex but highly logical structure that enables its users to be extremely clear and precise. Verbs serve as a good example. English has an elaborate verb tense, aspect and mood structure that permits writers to convey the relative chronology of actions with great precision. Expressing whether an action occurred before, after, or at the same time as another action can be crucial to your explanation of a problem. The accurate use of grammatical and syntactic features that produce precision is an important part of good technical writing.

If you are not confident about your grammar and syntax, refer to **Appendix B** at the end of this book for some basic instruction, and if you need more help, go to a university handbook such as *Harbrace College Handbook* or one of its competitors.

Diction

A matter related to, but distinct from, grammar is diction, or the way you choose to say things. All native speakers of a language know that there are ways that things are said (idiomatic ways). Equally, there are ways in which things are not said (unidiomatic ways). These unidiomatic constructions, though they may not *technically* violate a rule of grammar, reduce the readability and smoothness of your writing. For example, if you write, "The pendulum, which when it swings through having been bestowed with energy through an applied motion, imparts this energy to the mechanism that it is connected to, is a simple kinetic device," you haven't really committed a blatant grammatical or syntactic fault, but you haven't expressed yourself very idiomatically or effectively. Why not say 'The pendulum is a simple kinetic device; when an applied force sets it into motion, it imparts energy to the mechanism it is attached to'? When you're writing, you should always ask yourself 'Would I really say that?' If the answer is no, rewrite your sentence until it is simple, clear, and idiomatic.

On the other hand, it's important to guard against confusing idiomatic expression with excess informality. The language of advertising, popular journalism, and popular entertainment is, for the most part, unsuitable for technical communication, because it is generally imprecise, often slangy, and frequently non-standard in grammar and syntax. Writers like Martin Gardner, Richard Dawkins, Lewis Thomas, and many others provide examples of idiomatic, elegantly written prose that you can use as models instead.

Many writers worry about sounding 'too simple;' readers of technical documents, however, are likelier to appreciate a straightforward style than to value unnecessary complexity.

Tone

Tone is the term we use to indicate the level of politeness of a communication. Perhaps more precisely, it's the term we use to refer to the attitude the writer takes to the reader. Common sense suggests that you be polite to a superior and polite and cordial to people over whom you have authority if you want their willing cooperation. Difficulties with tone crop up often in documents produced by inexperienced writers: rudeness and imperiousness are common even in documents directed to managers. Usually the reason is that the writers fail to realize that politeness in speech is often communicated by vocal qualities that are absent from writing. A request uttered in a polite vocal tone and inflections may seem harsh when it's written down—that is, something as simple as "Come here now" may be *spoken* politely, but written it looks bossy.

Writers can achieve a polite tone in a number of ways. One of the easiest is to use conditional or subjunctive verb forms ("may *or* might come", "would help", "could be done") instead of indicative forms ("comes", "will help", "will be done "). These forms (also called 'modals') are linguistically complicated, but if you think about it, you can probably get a feeling for their effects: modals (the ones in the first set) leave room for the reader to judge, consider, or decide, while the indicative forms in the second set imply that what the writer says is true or must happen.

Another common way of achieving a polite tone is to phrase things as direct or indirect questions rather than as statements. "Will this plan really work?" or "I wonder whether this plan is going to work" is likely to strike the reader as more polite or deferential than "This plan won't work" or, even worse, "This plan is bad; it won't work." It's particularly important to avoid strings of imperatives (direct commands). "I wonder whether you could stop by my office this afternoon; are you available at 3:30?" has a much better tone than "Stop by my office this afternoon at 3:30," though if your superior says it to you, you're going to understand it as an order rather than as a real question. Nonetheless, people respond better to politely worded communications, generally, than to communications couched (inadvertently or deliberately) in what they perceive as a demanding, arrogant, or insensitive tone.

Tone is a product of many aspects of your writing working together, and it's not easy to identify the roles of individual elements, but you should always start, at least, by asking yourself what your relationship to your reader is. Then see whether your language—particularly the use of pronouns and the choice of verb forms—seems to reflect that relationship in a way that will be satisfactory to both of you.

Level

'Level' is the term we use to describe formality. To a large degree, level depends on the occasion that calls for the writing. A report to the board of directors of your company should be more formal than a memorandum to your immediate superior (unless he or she is extremely formal in manner and likes formality in writing as well; there's that concept of "audience" again.) Many inexperienced writers, however, respond to a requirement for formality by producing extremely abstract language. For example, in a formal setting, some writers resort to using phrases like "personal protection device" rather than saying "pistol". But the words could just as easily indicate a knife, a rolling pin, or an automobile. Confusing abstraction with formality results in writing that most readers will find hard to use.

Writers achieve formality primarily through their choice of words and their sentence structures. You may need longer sentences and more complex vocabulary than you would use in a casual setting. Typically, formal writing contains more one-word verb forms like 'extricate' and fewer two-word verb phrases like 'take out,' and it often displays infrequent use of personal pronouns like 'I' and 'you,' which establish a direct link between the writer and the reader. Passive sentences, which are impersonal in their subjects, are more common in formal style than in informal writing, and so are sentences with several clauses attached to the main clause. But don't confuse either abstraction or complication with formality. Maintain the goals of precision and clarity. You'll find it easier to write, and your readers will appreciate it.

Punctuation and Graphic Conventions

Punctuation and graphic conventions are extremely important aids to understanding; that's why English (and most other languages) developed them. Punctuating properly is a great service to the reader and in the some instances is the only way to save a given sentence from ambiguity. Many people think that the rules of punctuation are unnecessarily detailed, arbitrary, or irrational. For the most part, they aren't. They're simple to grasp and easy to learn if you take the time to understand the basic rules of English sentence construction. It is not the purpose of this book to teach punctuation, but **Appendix B** has some guidance, general principles, and tips for avoiding the most troublesome punctuation errors.

1.5 GRAPHICS IN TECHNICAL REPORTS

Most technical reports, particularly oral reports, contain graphics (also called 'visual aids' in the context of oral presentations). Often these are the first things a casual reader of your report notices. Because they are both highly noticeable and extremely useful, you need to pay close attention to the way you choose, design, and incorporate your graphics.

Some general considerations need to precede specific guidelines on graphics. First, it's important to recognize the two main purposes of graphics: (1) to make a visual impact and (2) to give information. If a graphic is designed for the second of these purposes, the information can be either structural (that is, it can inform the audience as to the parts of a presentation or communicate which part the speaker is moving into, for example) or substantive. If a graphic is intended to give substantive information, the designer needs to consider what kind of data makes up that information. And this is where the writer has to make some important decisions.

So the first general principle is this: *the nature of the information you want to convey dictates the kind of graphic you should choose.*

The second principle is simple but very important: in technical reports, *graphics **support** technical material, but they don't **replace** it.* They can be powerful tools in your overall argument, but by themselves they can't make an argument; that requires verbal reinforcement and explanation.

The third principle is another simple one: *graphics are almost never self-evident.* They almost invariably require titles or captions, and if they are images (as opposed to charts or graphs) they almost invariably require labels.

Guidelines for the design of graphics appear in **Appendix C.**

1.6 FORMATS FOR TECHNICAL REPORTS

Using conventional formats for your memoranda and reports is extremely important, and so is taking the time to make your documents look sharp. The reasons are these:

- Using conventional formats produces standardization of documents within an institutional setting, and standardization is helpful to users of the documents;
- Typically, the conventions themselves are established to simplify the reader's task;

- Using the format established for a particular institution allows you to demonstrate that you belong to the discourse community of that institution. This automatically separates you from non-members and gives you the presumption of credibility.

That said, you need to recognize that formatting conventions will vary from setting to setting. So in this task, as in many other engineering tasks, you'll need to work from detailed specifications that you receive and produce documents that conform to those specifications.

Before you look at the guidelines below, you should think about format once again in light of the statements above. It may help you if we isolate and explain the elements and their broad functions:

- Elements that provide information about the origin and occasion of the document and the history of its readership—i.e., indications of the writer and his/her position, the intended reader(s) and his/her/their position(s), the date, and the purpose;
- Elements that provide structural information—i.e., section headings, figure numbers, page numbers, tables of contents, lists of figures, spacing;
- Elements that provide hierarchical information—i.e., different fonts and font sizes, indentation.

Additional format specifications primarily arise out of consideration for (1) ease of reading—overall font size and margins, placement of text on the page, paper and print quality, particularly, or (2) ease of handling—paper size, the type of fastening or binding.

A document that you have taken care to produce according to specifications and that you have made as neat and attractive as possible gives the impression of professionalism.

For examples of formats in use, see the sections on individual types of technical documents in the main text of this book. Consult with your instructor or supervisor for specifications of the format in use in your organization or class.

2

BUSINESS MEMORANDA
AND SHORT REPORTS

The first piece of technical writing that we will present is the business memorandum, for two reasons. First, it exhibits most of the features found in other forms of technical writing. Second, it is a short and common form of technical writing. You should learn how to write memoranda properly, first, because you will undoubtedly need to write them early in your employment or school career and, second, because the principles we explain in this chapter are fundamental and often underlie other types of technical reports.

Purpose of the Business Memorandum

This document is the most basic form of internal communication (unlike a letter, which goes to someone outside of your company or organization). Therefore, it may be used for any communicative purpose, and *thus it may be either informative or persuasive.* An informative memo may simply communicate that you have completed a task or an investigation and give the results that you obtained, or its purpose may be to point out a problem that needs attention. If its purpose is persuasive, the memo may ask permission to take action or it may justify actions you have already taken. It is often useful to tell your reader explicitly what the purpose of the memo is.

Audience for a Business Memorandum

Because a business memo is such a basic form of communication and may have any of a number of different purposes, it may have any of a number of different audiences. By "audience" we mean the intended reader. *It is important to realize at the outset that any piece of technical writing will have a "primary audience"—the person to whom the memorandum is addressed and primarily intended, and a "secondary audience"—any other readers who may be given the memo.*

The primary audience of a business memo might be your manager (perhaps you are reporting on work you have accomplished for him or her) or it might be co-workers or employees (for example, if you are announcing or explaining a program or policy).

The secondary audience for your memorandum might be those to whom the primary recipient might give a copy should he or she need advice or opinions. These might be accountants or perhaps lawyers if there are economic or legal consequences to what you are reporting or proposing. Understand that your secondary audience is unlikely to have the same interest in the memo (or knowledge of its subject matter) that the primary audience does, and it is useful to keep that in mind, particularly if you know that there will be a secondary audience and you wish to influence it.

Overview for a Business Memorandum

A memo often begins with an overview made up of two brief sections that orient the reader and tell him or her what to expect (or whether it is important to read this particular memorandum at all). It is useful to call these sections the "foreword" and "summary," although you may see memos with only what is called a "summary." The foreword is meant to give the reader the context of the memo and the Summary generally states what actions you have taken and their results. Very short memoranda may have nothing more than a simple introductory section (see Chapter 1).

Main Sections

The typical memo includes an overview section (commonly comprising a foreword and summary) and then a body section (often entitled "Discussion" or "Details") that gives the reader all of the information your memorandum needs to convey. The body includes a separate introduction and subsections. Some possible modes of development for the paragraphs contained in the body are discussed in **Appendix B;** the most typical modes for technical reports are amplification/extended definition, comparison/contrast, cause/effect, process description, and physical description. Chronological development or narration is also common in background sections.

Other Materials

Memos occasionally need supporting documents, graphics or citations, although these are more typical of some other types of technical reports. See the appendices for help with these elements if you need to include them.

Format

Memoranda have conventional formats. If you are a student, your instructor will supply you with guidelines; if you are employed, ask your supervisor about obtaining a company style sheet or suitable model.

MEMORANDUM TEMPLATE

TO:

FROM:

SUBJ:

DATE: (may also be placed at the top, depending on your organization's preferences)

ATTACHMENTS: (if applicable)

DIST: (if applicable)

FOREWORD

Statement of the problem or issue that faces your organization or department

Statement telling why the problem or issue is important

Statement explaining what your task or role was with regard to the problem or issue

Statement telling what you did in response to the task you were assigned

Statement telling what the memo contains and its purpose

SUMMARY

Presentation of the main findings or results you obtained in the course of responding to your task (if you were asked to provide a recommendation, this usually comes first)

Brief statement of the support for your main findings or results—often a brief explanation of how you arrived at your findings or results

Statement of managerial implications, if applicable

DETAILS

This section of the memorandum sets out in detail what you have summarized in the foreword and summary. While the foreword and summary may amount to only half a page, your details may take up two, three or more. Consider putting in subheadings; they are generally very helpful to readers.

DOCUMENTATION (if needed)

1ST EXAMPLE OF MEMORANDUM

WOLVERINE ELECTRONICS, INC.
1817 Victors Way
Ann Arbor, MI 48104

To: Dr. Dwight Stevenson, Manager, Facilities

From: M. A. Gilbert, Project Engineer

Subject: Proposed lighting for Terre Haute facility

Date: 12 September 2006

Dist: Bob Brown, Manager, Wolverine/Terre Haute

Foreword

Last year our company bought a small manufacturing plant in Terre Haute, Indiana, formerly used for making brake drums for the automobile industry. It is in the process of conversion to a facility for making our ZX-series transistors for various electronics firms. The existing lighting, while adequate for work on automotive parts, is not adequate for the new work. Therefore, I was asked to find the best lighting for the plant. I conducted research and identified the applicable regulations. This report gives my findings and recommendations.

Summary

My research into trade journals and standard texts on lighting, in combination with the regulations given in OSHA publications, shows that it is necessary for the shop floor to be exposed to μ lumens at any given point. Furthermore, full spectrum light is necessary in order for workers to be able to spot defects in the ceramic coatings of the transistors. Further research indicates that 220 fluorescent lights, each of μ watts, are necessary to illuminate the shop floor. The two most economical solutions I discovered are to

- convert all of the plant light fixtures to accept Durabright fluorescent lights at a cost of $35,000 or
- convert to Zeta fluorescent lights at a cost of $20,000.

Although the Durabright lights cost $25 each and therefore are more expensive to install and replace, they last 30% longer and use 50% less power. In turn, the smaller power-draw will allow us to continue using the existing circuit breakers in the plant. Thus, though Durabright lights are the more expensive of the two, installation of them coupled with the continuing use of the existing circuit breakers will result in savings over the Zeta lights after eight months of use. Accordingly, I recommend that we convert the plant to the use of Durabright lights.

Discussion

Current Situation

Last year our company purchased a brake drum manufacturing plant in Terre Haute, Indiana. We are currently converting it for use to produce transistors for various radio manufacturers. We have discovered that the lighting, which was adequate for work on automotive parts, will not be sufficient for the new work. In particular, we will rely on trained workers to visually spot defects in the transistors, and inadequate light may lead to an unacceptable error-rate in spotting defects. Therefore, Bob Brown, manager of the Terre Haute plant, has asked me to determine the adequate lighting for the plant. . . **[The paragraph should continue with further details.]**

Research

I conducted research in the trade journals *Lighting Today, Industrial Illumination* and *Photonics and Industry.* I went further by getting information from standard textbooks on lighting and from OSHA publications No. 3387-6 and 3288-9, which give the bureaucratic regulations for lighting in the workplace. The research showed that it is necessary for the shopfloor to be exposed to μ lumens . . . **[The paragraph should continue with further details.]**

In the course of my research I looked into converting to lights made by five different manufacturers, Durabright, Zeta, Wilson, Extrovert and Limelight. Each of these lights has distinct characteristics . . . **[Several paragraphs should continue with further details.]**

Alternative Solutions

In the end, I have decided that only two of the five present us with acceptable choices: Durabright and Zeta. It would cost us $35,000 to convert the plant's lights to Durabright lights and $20,000 to convert to Zeta lights. However, the installation cost is not the only thing to consider; we must consider also the life of the lights, the power they draw, and whether we will be able to continue using the existing circuit breakers . . . **[Several paragraphs should continue with further details.]**

Proposed Solution

Therefore, even though Durabright lights are the more expensive of the two, installation of them coupled with the continuing use of the existing circuit breakers will result in savings over the Zeta lights after only eight months of use. Accordingly, I recommend that we install 200 Durabright lights at a cost of $35,000.

2ND EXAMPLE OF A MEMORANDUM

LETTERHEAD

To: Ms. Mercedes Barros, Supervisor

From: Colleen Budd, Engineer

Date: April 5, 2008

Subject: Estimate of the electrical generating potential, dam height, and reservoir size for Rio Consuelo site

Encl.:   Calculations for the potential power generation

Foreword

Our company has recently acquired a new dam site, and I was asked to perform a preliminary estimate of the electrical generating potential of the site. I analyzed the site location, evaluated the constraints and completed some calculations. This report includes my findings concerning the constraints that will influence the dam height. This report also includes my recommendations for dam height and estimates of both the hydroelectrical generating potential and the resulting reservoir size.

Summary

The upstream elevation and various constraints were used to determine the recommendation for the height of the dam. On the basis of a detailed evaluation of the Rio Consuelo topographic maps, I recommend a dam height of 50 meters for the new site at 45° 28′ S, 72° 18′55″ W. A dam of 50 meters will create a reservoir approximately 1.2 kilometers in length. With the average flow rate of 174m^3/s and an assumed efficiency of 85%, the hydro-electrical generating potential is 72.5 Megawatts. . . **[The paragraph continues with further details.]**

Details

After studying several topographic maps of the Rio Consuelo area, I have determined the upstream elevation profile of the site. The elevation of the proposed dam site at 45° 28′ S, 72° 18′55″ W is approximately 120 meters above sea level. Roughly 4 kilometers upstream in the east-southeast direction, the Rio Torro joins the Rio Consuelo. Approximately 17 kilometers southeast of that point, on the eastern bank of the Rio Simpson, is the city of Coihaique, the capital of Region 11 of Chile. The elevation of the river at this location is 215 meters above sea level, but the elevation of the city varies from 275 meters near the river to 375 meters further east. Many fields and farms surround the river in the vicinity of the city.

Because of the upstream profile of the dam site, I have determined three constraints that will influence the height of the dam. First, because the Rio Torro flows into the Rio Consuelo, there may be increased areas of inundation at the mouth, especially in the spring when the snow from the mountains melts and causes an increase in the rivers' flow rates. Second, the slope of the land becomes very flat approximately 10 kilometers upstream from the mouth of the two rivers. This relatively flat land is where Coihaique and the surrounding farms are situated. This level land is not suitable to contain a reservoir. Third, a main highway, CH 245, runs along the length of the Rio Consuelo from Coihaique to beyond the dam site to the city of Puerto Aisen. This highway is the connector between these two major cities. Also, the environmental effect of the dam and reservoir must be considered. The Rio Consuelo is known for fly-fishing (1), and the reservoir will cause a change in the ecosystem for the fish and other wildlife. A higher dam will result in a larger reservoir, which will consequently increase the population of wildlife that will be forced to adapt or die.

After taking into account each of the constraints, I recommend a dam height of 50 meters. The contour lines on Map 1 represent an elevation difference of 50 meters. The lowest contour line crosses the Rio Consuelo 1 kilometer upstream from the dam site. Therefore, the length of the reservoir will be approximately 1.2 kilometers. If the dam were to be any higher than 50 meters, the reservoir size would become much greater because of the leveling of the leveled slope of the land further upstream. For instance, if the dam height were to be 100 meters, the reservoir size would extend to the second contour line in Map 1. This increase in reservoir area would result in the flooding of a vast area of more than 12 kilometers (from Map 1), potentially including the farms and city of Coihaique. Furthermore, as the height of the dam increases, the length of CH 245 that will be flooded also increases. Building a 50m dam will require that only 1 kilometer of CH 245 will need to be rebuilt at a higher elevation, as opposed to at least 12 kilometers with a 100 meter dam.

Therefore, I recommend a dam height of 50 meters for the new site at 45° 28′ S, 72° 18′55″ W. A dam of 50 meters will create a reservoir approximately 1.2 kilometers in length. With the average flow rate of 174m^3/s and an assumed efficiency of 85%, the hydro-electrical generating potential is 72.5 Megawatts.

[References should be included at the end of the memorandum body.]

[Calculations should be attached to the memorandum.]

EXAMPLE OF A SHORT REPORT
IN MEMORANDUM FORM

MEMORANDUM

Date: November 8, 2007

To: Dr. Abe C. Dee
           Director, Biomedical Engineering Resource Center

From: The Back Pack: [Name 1, Name 2, Name 3, Name 4]
           Interns

Subject: The pedicle screw system as a treatment for adolescent idio-
           pathic scoliosis: description and evaluation

Foreword

Adolescent idiopathic scoliosis, a spinal disorder causing abnormal curvature and rotation, affects as much as 4% of the U.S. population between
the ages of 10 and 18. Because the condition is so widespread and
because its consequences can be severe, biomedical engineers have created numerous orthopedic devices to correct it. You asked us to examine
and report on one of these devices, and we selected the pedicle screw system. To carry out this task, we researched the medical problem of scoliosis and the anatomical structures it affects, the history of treatments for
scoliosis, and the current pedicle screw design. The purpose of this report
is to give you our findings.

Summary

The pedicle screw system, used in conjunction with the surgical process
of spinal fusion, is an effective treatment for adolescent idiopathic scoliosis, or abnormal curvature of the spine. By realigning and stabilizing
the spine, the implant corrects the deformity and prevents further progression of the scoliotic curve.

The spine contains 24 vertebrae in its three largest regions, the cervical
spine (the neck), the thoracic spine (the chest area), and the lumbar spine
(the lower back). Each vertebra is composed of a vertebral body, pedicles, and lamina; stacked together, these form the spinal canal. They surround and protect the spinal cord. Soft-tissue structures called intervertebral discs lie between pairs of vertebrae; these provide flexibility and act
as shock absorbers. Joints called facet joints connect the spinal segments

and permit a fairly wide range of motion. The spine is capable of rotational and translational movement about the x, y, and z-axes.

Scoliosis is a disorder characterized by an abnormal curvature of the spine. Three types occur: idiopathic, congenital (caused by birth defects), and neuromuscular (caused by diseases such as muscular dystrophy). The cause of the most common, adolescent idiopathic scoliosis, is unknown. It is defined as a spinal curve of more than 10° (measured by the technique called the Cobb angle) and is usually accompanied by a rotation of the vertebra. For curves of 40° or more, orthopedists often recommend surgery to correct the problem before the curve progresses and causes health complications, such as impairment of lung function.

Early attempts to treat scoliosis involved external bracing and manipulation of the spine. These methods were ineffective, because the spine reverted to its distorted position as soon as it was no longer supported. In the early part of the 20th century, orthopedists began using a technique called spinal fusion to fix the spine in place with bone grafts between vertebrae; by itself, however, fusion was not strong enough to prevent the progression of the curve. An orthopedic implant called the Harrington Rod, introduced in 1947, was the first major device designed to supplement spinal fusion in the correction of scoliosis. The Harrington Rod system could not correct rotational deformity of the spine, however, and because of its inflexibility it also typically led to further problems. Throughout the following decades, engineers created other forms of spinal instrumentation, including the Luque and Cotrel-Dubousset systems. These were more successful in the treatment of spinal rotation, but both had numerous complications. The development of the pedicle screw system in recent years represents a significant advance over these earlier systems.

The pedicle system employs pairs of titanium screws fixed into the pedicles of affected vertebrae. The screws secure short titanium rods that the orthopedic surgeon places and bends to straighten and de-rotate the spine. The rods and pedicle screws support and stabilize the spine while fusion of one or more spinal segments, initiated by a bone graft, occurs.

During the implantation process, the surgeon first removes the intervertebral discs to make the scoliotic curve flexible. He or she next drills threaded holes into each pedicle and inserts the screws; he or she then places rods to connect the pairs of screws and applies torque to the spine to twist the curve to the appropriate angle. Finally, the surgeon inserts bone graft into each intervertebral gap to initiate fusion.

The function of the pedicle screw system is essentially to counter the bending forces that cause the scoliotic curve to progress. The system does this by applying a constant force to the convex side of the curve. As the spine straightens, the bending moment decreases. As a result, the fusion process can take place and, when complete, can hold the spine in its corrected position.

The pedicle screw system remains in the body as a support for the spine during and after fusion. Most patients are able to return to normal moderate activity within two weeks after the surgery, and few experience significant complications.

Introduction

Scoliosis, a disorder of the spine characterized by abnormal curvature, is a very common medical problem, especially among young people. While it is not usually severe enough to require significant treatment, approximately two to four percent of all persons between the ages of 10 and 18 experience a form of the condition called adolescent idiopathic scoliosis, defined as an abnormal curvature of the spine greater than 10° (citation, date). These patients suffer discomfort and are often unable to carry out everyday activities. Severe cases can lead to serious and debilitating cardiovascular and muscular problems (citation, date). For these reasons, orthopedists have devised numerous forms of spinal instrumentation to treat adolescent idiopathic scoliosis and reverse its effects. One of these, the pedicle screw system, used in conjunction with spinal fusion, works well to correct the abnormal curvature and prevent further progression. It counteracts the forces causing the abnormal curve and supports the vertebra during the fusion process (citation, date).

This report discusses the pedicle screw system in detail. It begins with a brief treatment of the anatomy of the spine and a description of scoliosis. It follows with a short historical survey of approaches to the treatment of scoliosis. The final sections give a physical description of the pedicle screw system and the implantation process and, finally, explain the mechanics and functioning of the system.

Structure of the Spine

Because scoliosis affects the spine, an understanding of its anatomy is essential. As Figure 1 shows, the spine contains 24 vertebrae in its three largest regions, the cervical spine (the neck), the thoracic spine (the chest area), and the lumbar spine (the lower back). Each vertebra is composed of a vertebral body, pedicles, and lamina; stacked together, these form the spinal canal. They surround and protect the spinal cord. Soft-tissue structures called intervertebral discs lie between pairs of vertebrae; these provide flexibility and act as shock absorbers. Joints called facet joints connect the spinal segments and permit a fairly wide range of motion. The spine is capable of rotational and translational movement about the x, y, and z-axes.

Figure 1: Regions and natural position of the spine (source)
[Note that captions may be placed above or below images, according to style guidelines for your organization or class. Images may be bordered or not, again according to the guidelines you receive.]

The spine has a natural s-shaped contour, as Figure 1 shows. This natural contour allows the spine to handle stress . . . **[The paragraph and the rest of the section would continue with further details about the pertinent anatomy, physiology, and kinematics of the region. Additional figures might also be included.]**

Scoliosis

Scoliosis is a disorder characterized by abnormal curvature and rotation of the spine (citation, date). Three types of scoliosis occur: congenital, neuromuscular, and idiopathic. Congenital scoliosis occurs in children from birth, usually as a result of malformations of the vertebrae. Neuro-muscular scoliosis results from neuromuscular disorders such as muscu-lar dystrophy and cerebral palsy. The etiology of the remaining type, idiopathic scoliosis, is unknown. It is the most common form of the dis-order, accounting for 80-85% of all cases (citation, date).

Orthopedists classify idiopathic scoliosis into three subtypes, according to the patient's age . . . **[The paragraph would continue with a discus-sion of these three types.]** This report focuses on adolescent idiopathic scoliosis (AIS), because this form of scoliosis is most commonly treated with pedicle screw systems (citation, date).

Diagnosis and Description of Adolescent Idiopathic Scoliosis

AIS is diagnosed when an adolescent displays a spinal curvature of more than 10 degrees, as Figure 2 shows. In the more severe forms of the dis-order, when spinal rotation accompanies the curvature, one of the patient's shoulders typically juts forward and the ribs protrude at the back on the opposite side. The greater the degree of the curvature and the greater the growth potential in the patient, the greater the chance of curve progression.

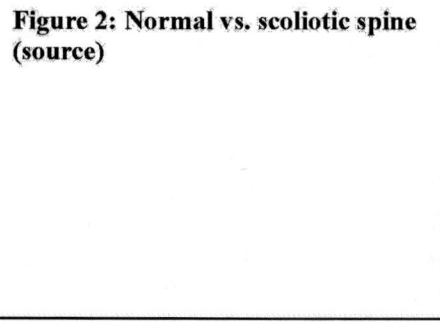

Figure 2: Normal vs. scoliotic spine (source)

Cobb Angle

To determine the degree of curvature in the spine, orthopedists use a tech-nique called measurement of the Cobb angle, as shown in Figure 3. To measure this angle, the doctor draws a straight line through the top of the vertebrae at the top end of the curve and through the bottom of the verte-brae at the bottom end of the curve . . . **[The paragraph would go on to** explain the method and discuss what various results imply for the patient.]

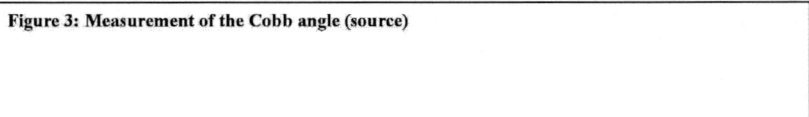

Figure 3: Measurement of the Cobb angle (source)

The Development of Treatments for Scoliosis

The earliest approaches to the treatment of scoliosis were external bracing and manipulation of the spine (citation, date). Primitive spinal surgeries are attested in the mid-eighteenth century, but no effective procedures really developed until the middle of the twentieth, with the advent of spinal fusion and spinal instrumentation . . . [The paragraph would continue with an introduction of the main systems that the section will discuss.]

The Harrington Rod System

One of the first forms of spinal instrumentation, introduced in 1953 by Dr. Paul Harrington, consisted of a stainless steel rod attached to the spine at two points on the concave side of the spinal curve (citation, date) . . . [The paragraph would continue with a short explanation of the system and its performance. An image of the system might be included, if space permits.]

The Luque System

In response to the numerous complications of the Harrington Rod system, Dr. Edward Luque devised a modifed rod system in 1976. The Luque System consists of two long, cylindrical rods fixed to the spine on two sides . . . [The paragraph would continue with a short explanation of the system and its performance. An image of the system might be included, if space permits.]

Cotrel-Dubousset Instrumentation

In the mid 1980s, a third form of scoliosis correction, Cotrel-Dubousset instrumentation, emerged. This system employs two parallel rods fixed by hooks to the spine . . . **[The paragraph would continue with a short explanation of the system and its performance. An image of the system might be included, if space permits.]**

[The section would conclude by bringing the historical survey up to the present and introducing the device that the remainder of the report will treat.]

The Pedicle Screw System

The pedicle screw system, used in conjunction with spinal fusion, corrects the deformity produced by AIS and allows the spine to maintain proper position during the patient's adolescence, when rapid growth would otherwise cause acceleration and worsening of the curve. To achieve this correction, the system employs a system of screws fixed into the pedicles of the affected vertebrae, attached rods that exert forces to counter the bending moment, and bone grafts to fuse the spine in its corrected position.

Components of the Pedicle Screw Systems

[This section would enumerate the components of the system and then give a physical description of each one, accompanied by images.]

Implantation of the Pedicle Screw System

[This section would explain the implantation process step by step.]

Function of the Pedicle Screw System

[This section would explain how the system achieves correction, with diagrams to illustrate the forces acting on the system.]

Conclusion [Optional]

The pedicle screw system is an effective approach to the treatment of AIS . . . **[this section would restate the problem and the solution briefly and, if appropriate, suggest further research directions.]**

References

[This section would list *all* sources cited or consulted, presented in the format your organization or instructor specifies.]

3

PROPOSALS

Purpose of a Proposal

Formal proposals are specialized technical business documents that offer work (services or products) that fulfills a need for a particular audience. The audience for the proposal can be internal to an organization, such as the management team or another department, or the audience can be a client who is external to the organization. In either case, a proposal needs to be both informative and persuasive because, often, you will need to convince your audience to decide to accept the services or products that you are proposing. To be convincing, a proposal should indicate the writer's understanding of the need and the criteria for successfully meeting that need. It should also explain the details of the work, how it will be completed, when it will be completed, who will complete it and how much it will cost.

Some proposals may be referred to as "informal proposals." Informal proposals are just shortened versions of formal proposals or may contain only some of the sections of a formal proposal.

Preparation for Writing a Proposal

Prior to writing a proposal, you should clarify the needs of the client and identify the details of the specific work that is required. The expected results and validation method for a successful project from the perspective of the client and user should be defined whenever possible. Sometimes the client may send a Request for Proposal (RFP) that provides some of the necessary details about the work or project, including particular requirements and deadlines. You can obtain more specific information about the requirements by interviewing or meeting with the client. Also, prior to writing a proposal, you should identify the available resources and expertise within your organization. Finally, you should collect information to define budgetary requirements and limitations.

Front Matter and Overview for a Proposal

Proposals begin with a cover page and a table of contents. The overview will typically be an executive summary or a foreword and summary. For these elements, see **Chapter 2**.

In an executive summary, you should provide an overview of the entire proposal, including the main issue or problem leading to the need for the proposed work, the main points of the proposed work, and the methods by which the proposed work could solve the main problem. You can also include a condensed version of the costs and major budget issues in the executive summary of a proposal. The executive summary may conclude with an organizational forecast statement. The forecast statement offers the reader a preview of the topics in the order that they appear in the body.

Main Body Sections

The body of a proposal begins with a simple introduction. In a simple introduction for a proposal you should include, at a minimum, the scope of the proposed work. You can define the scope of the proposed work by explaining what the proposed work is and what it is not. You can define the scope of the proposed work in relationship to other work that has previously been conducted or that is currently being conducted. By defining the scope, you are able to effectively focus the audience on the primary purpose of the proposal. In addition to defining the scope, you may also include information to help the audience understand the current need for the proposed work.

The remaining sections of the body of a proposal may include any of these elements:

1. **Background information** that may help the audience to better understand the details of the proposal or information related to the project that may place the proposal in context with other work.
2. **Details of proposed work** explaining specific tasks and including an explanation of how the work meets the required objectives, how the work will be completed, and when the work will be completed. This section can include alternative solutions and an explanation of why the alternative solutions are not recommended. By discussing alternative solutions, you show that you have a thorough understanding of the requirements and possible solutions.
3. **Explanation of the validation, testing or evaluation** of the proposed work.
4. **Material, personnel and equipment requirements** detailing all of the requirements, including a description of the coordination and management of personnel and sub-contractors, if necessary.
5. **Explanation of the expertise**, reputation, experience and success of your organization as these relate to the proposed work.
6. **Detailed budget with costs** (often in table format).

Closing Section
In the "Conclusion" section for the proposal, you should clearly restate your main recommendations for the proposed work and reiterate how, on the basis of the writer's understanding of the original problem and the criteria the writer has articulated, the proposed work clearly satisfies the requirements.

Other Materials

Proposals often include an appendix, which may include supporting documentation such as

1. Details of similar projects your team or organization has completed
2. Detailed qualifications of personnel (such as a resumes) and subcontractor expertise
3. Letters of recommendation from other client(s).

See Appendices **A, B, C, and E** for content development, language usage, graphics and evaluation checklists.

PROPOSAL TEMPLATE

COVER PAGE

OVERVIEW: EXECUTIVE SUMMARY OR FOREWORD AND SUMMARY

SIMPLE INTRODUCTION INCLUDING SCOPE OF PROPOSED WORK

BACKGROUND

Information the reader needs in order to understand the proposal
Information demonstrating that the proposal writer understands what is involved

DETAILS OF PROPOSED WORK

Clear explanation of the tasks involved in accomplishing the proposed work and a statement of how long it will take to accomplish them

EXPLANATION OF VALIDATION, TESTING OR EVALUATION OF PROPOSED WORK

Discussion of how successful completion of the work will be judged

MATERIAL, PERSONNEL AND EQUIPMENT REQUIREMENTS FOR PROPOSED WORK

Statement of what is required to accomplish the proposed work, such as materials, manpower and equipment

EXPLANATION OF THE PROPOSER'S EXPERTISE

Statement of the qualifications of those who propose to do the work, including those of any subcontractors

DETAILED BUDGET

Clear statement of the costs involved

Costs presented in table form

CONCLUSION

Statement of the main recommendations of the proposed work

Reiteration of how the proposed work meets the criteria or other requirements established by the proposal writer or the client

REFERENCES (if needed)

APPENDICES (if needed)

EXAMPLE OF PROPOSAL

ARSENAL SHIP MECHANICAL
AND ELECTRICAL SYSTEMS DESIGN PROJECT

USN Proposal Number 3277-889

Submitted by:

Marcia Wilson, Engineering Project Manager
U.S. Ship Design, Inc.
1298 Radway
Virginia Beach, VA

Date of Submission:

February 3, 2008

Submitted to:

Dr. Aliya Jones, U.S. Navy
Dr. Ariana James, NAVPYS Corporation

Executive Summary

U.S. Ship Design has been asked to propose a reliable and redundant design for the mechanical and electrical systems for arsenal ships. The design team at U.S. Ship Design has evaluated various engine options and electrical configurations and in this document proposes an economical, low-maintenance and functional design for arsenal ships.

An arsenal ship is designed to more readily survive combat than ships of conventional design; features contributing to its hardiness include a reduced radar cross section, a double hull design around the missile compartments, and subdivision of the hull. The use of standardized parts on the ship allows effective production methods and results in the reduction of the overall cost of the vessel.

The proposed mechanical system for arsenal ships, which will meet all requirements for speed and redundancy, consists of an electrical drive system and two gas turbine prime movers, one General Electric LM 2500 and one General Electric LM 1600. Two Siemens and Schottel twin-screw podded propulsion systems are proposed because they would take less space in the engine room than a conventional propulsion system and would improve maneuverability and efficiency. All components of the main engine support system and auxiliary machinery are to be doubled to increase the chances of the ship's survival in combat.

The electricity produced by the electrical generators would be distributed throughout the ship and converted into different forms of energy through power conversion modules. In addition, a Pratt and Whitney ST6-79 generator would be available to power emergency systems in case the main system were to become inoperable.

The cost to build both the mechanical and electrical systems (including both material and labor) totals approximately $31,100,000, plus 6% overhead and a 12% profit margin. The cost to design and draft the mechanical and electrical systems over a three-month period totals $129,000.

This report includes a description of the proposed mechanical and electrical systems, including the electrical requirements, testing and analysis requirements, and emergency systems. It also includes a list of personnel requirements, a table of costs, and an overall timeline.

1.0 Introduction

An arsenal ship is designed to survive attacks; features contributing to its hardiness include a reduced radar cross-section, a double hull design around the missile compartments, and subdivision of the hull. Its mechanical and electrical systems must reliably and efficiently propel and power the ship. This document presents a description of the mechanical and electrical systems design components we propose for arsenal ships; it also includes a cost estimate and timeline for the design.

2.0 Details of Mechanical and Electrical Systems

This section treats the advantages of an electrical drive system over a geared drive system and discusses the engine and propellers that should be used with it.

2.1 Geared vs. Electrical Drive Systems

On the basis of comparisons of cost, maintenance, and size, we recommend the less conventional electric drive system for arsenal ships. An electrical drive costs 12% more overall than a geared drive system but costs 23% less to install. It requires 47% fewer spare parts and 46% fewer maintenance hours. On the basis of this data we conclude . . . **[Paragraph continues with further details.]**

Use of an electrical drive system simplifies system maintenance because. . . **[Paragraph continues with further details.]**

An additional advantage of an electrical drive system is that it employs no shaft, so the engine room can be placed either forward or aft of the missile bays without fear of interference between the missiles and a shaft. Last, because ship service loads and propulsion are coupled in an integrated system, the prime movers operate at or near rated speeds . . . **[Paragraph continues with further details.]**

```
┌─────────────────────────────────────────────────────────┐
│                                                         │
│                                                         │
│              (Table–comparison of engines)              │
│                                                         │
│                                                         │
└─────────────────────────────────────────────────────────┘
```

Table 1: Efficiency, output, and fuel economy figures for GM LM 2500, GM LM 1600, Mitsubishi X-2500 and Daewoo 1600-Z

2.2 Main Engine Selection

We propose a combination of gas turbine and gas turbine prime movers. The LM 2500 is rated at 25,000 kW; a single one, therefore, does not meet the ship's need for 34,000 kW of power. Adding another LM 2500, however, would produce too much power. Therefore, GE's smaller engine, the LM 1600, will be used as the secondary engine. The LM1600 is rated at 14,920 kW, so it can handle the ship service loads and the propulsive load up to 4 knots in case the LM 2500 becomes inoperable.

These two engines were compared with similar engines produced by two other manufacturers; Table 1 shows fuel economy of the four engines considered and power ratings for each . . . **[Paragraph continues with further details.]**

2.3 Propeller Selection

In addition to the requirements for speed and redundancy, two main goals guided the selection of the propeller: decreasing the size of the engine room and improving maneuverability. The use of a podded propulsion unit permits a smaller engine room and improves flow to the propellers; therefore, we propose that a podded unit be selected. After evaluating several units in regard to their cost, reliability, and electrical requirements, we decided . . . **[Paragraph continues with further details.]**

2.4 Electrical Requirements

Each unit contains its own electric generator (see Figure 1 above). The electric generators produce 6120 VAC, which is used directly by the propulsion motors. Power conversion modules convert the remainder of

(Graphic–electrical system of propulsion units)

Figure 1: Schematic of Siemens and Schottel twin-screw podded propulsion system (source)

the power to voltages that can be used throughout the ship . . . **[Paragraph should continue with further details.]**

2.5 Emergency Systems
In the event of failure in the engine room, certain systems must continue to run. These include . . . **[Paragraph should continue with further details.]**

3.0 Validation, Testing and Analysis
The mechanical system is a configuration of pre-existing parts that have already undergone testing. Test results . . . **[Paragraph should continue with further details.]**

4.0 Cost Estimation
The total weight of material used for the machinery and electrical designs and the estimated cost of labor and materials for both systems are summarized in Table 2. The required overhead cost is 6% of the total cost. A profit margin of 12% has also been added . . . **[Paragraph should continue with further details.]**

(Table–estimated costs)

Table 2: Estimated cost to build the machinery and electrical components for arsenal ships

5.0 Estimated Project Timeline

The design of the mechanical and electrical systems can be completed in 3 months (Figure 2). The basic plans for the mechanical and electrical systems will be tested thoroughly in an effort to eliminate all flaws early in the process . . . **[Paragraph should continue with further details.]**

> (Graphic of timeline for project)

Figure 2: General timeline of completion of the mechanical and electrical systems over the three-month period

In the second month of the design process, the final design of both the machinery and electrical systems . . . **[Paragraph should continue with further details.]**

In the third month, the mechanical and electrical systems will be drafted and . . . **[Paragraph should continue with further details.]**

6.0 Qualifications of Personnel

All personnel at U.S. Ship Design who will be involved in the design have qualifications appropriate for the tasks that will be assigned to them . . . **[This paragraph would continue with the names, credentials and experience of the principal members of the design team, possibly in table form. Resumes might be added as an attachment.]**

7.0 Budget for Design Work

Completing a functional design in three months will require that nine in-house staff members be assigned to the project. A director for the project will work 4 hours per workday to keep the design project running smoothly, and a clerk will work 2 hours/day to perform minor tasks. Six full time engineers and two full time drafters will design, draft, and test the systems for the arsenal ship. . . . **[Paragraph should continue with further details.]** Table 3 summarizes costs involved over the three-month period.

```
(Table–budget figures)
```

Table 3: Estimated design budget for systems

8.0 Conclusion

The proposed mechanical and electrical systems meet the objectives for speed and redundancy specified for the arsenal ship design.

The proposed mechanical system consists of an electrical drive system and two gas turbine prime movers, one General Electric LM 2500 and one General Electric LM 1600. Two Siemens and Schottel twin-screw podded propulsion systems are used to increase space in the engine room and improve maneuverability and efficiency. All components of the main engine support system and auxiliary machinery are doubled to increase the chance of survival . . . **[Paragraph should continue with further details.]**

The electricity produced by the electrical generators is distributed throughout the ship and converted into different forms of energy through power conversion modules. In addition, a Pratt and Whitney T6 generator is available to power emergency systems in case of an engine room casualty . . . **[Paragraph should continue with further details.]**

The cost to build both the mechanical and electrical systems (including both material and labor) totals approximately $1,100,500, plus 6% overhead and a 12% profit margin. The cost to design and draft the mechanical and electrical systems over a three-month period totals $ . . . **[Paragraph should continue with further details.]**

References

Appendices with relevant material may follow as necessary.

4

PROGRESS REPORTS

Purpose of a Progress Report

You write this document to tell those for whom you are doing a project or task how far you have come with the job and whether you will finish it on time. This purpose means a number of things, first among them that for your audience to understand what you tell them, they must know in some detail what the project is. Therefore, if this is the first progress report you are doing for a project, *you must describe the project adequately before telling the reader how far you have gone with it.* A second or third progress report in a series may not require a thorough description of the project; you may refer the reader to the first report.

Most projects of any size will be made up of different tasks, for example research, design, testing, re-design and so forth. This circumstance leads to a second important point: the information in a progress report must be presented in a way that makes it as easy as possible to understand. *As a rule, this means that the information you give should be organized topically rather than chronologically.* At first glance it might seem most natural to simply describe what you have done as a sequence of events: "First we researched this, then we did a preliminary design, then we researched that while we revised our design, and then we tested this other thing . . . " and so on. However, because most projects involve tasks which are of different kinds, the mere fact that they must be done one after the other (or at the same time) doesn't mean that this sort of arrangement in the report will make the information at all clear to the reader. Usually, the reader will better grasp the progress you have made if you organize your report by topic or category.

Therefore a good technical report will be organized on these general lines: there will be a description of the project first, including its aims, scope and duration. This information is all very necessary if the reader is to follow what comes after. This first section may be called the introduction or background. Then follows the body of the report, which should tell the reader how far you have come with the project and what remains to be done. It is best not to do this by simply listing the tasks that make up the project and stating that this one is done and that one is not and the next one is not done either but the one following is. It stands to reason that such a scattershot approach will make it impossible for the reader to keep in mind what's being asserted and to take it in as a whole.

Organization of the body of the report by category suggests this natural structure:

1. tasks completed
2. tasks not yet completed.

The tasks in the description of the project found in the introduction will be found under either of these two sections and, this way, the reader can grasp quickly the status of the project. Further, you should state when each completed task was done and when each uncompleted task will be done. Other organizations by topic are possible and perhaps preferable in some cases. The main point is this: organizing your material by topic will make it easier for a reader to grasp than telling what you've been up to chronologically as though you were telling a story. Within each of these sections you should deal with the tasks in question in the order that is easiest for the reader to follow.

Audience for a Progress Report

Generally you write a progress report for a manager who needs to know how far you have come along on a project that you or your team is working on. You should keep this in mind when considering the tone you use in addressing the reader.

Overview for a Progress Report

Like any other report, a progress report may begin with an overview of some sort. (See Chapter 1.)

Main Body Sections
Introduction

A progress report will begin with an introduction. This introduction often includes a forecast—a statement telling the reader what will be in the report. The forecast is followed by a description of the project itself so that the reader can understand the rest of the report. Be aware that an adequate description of your project may include a brief description of the problem that your project seeks to address. As always in technical writing, you must give sufficient context to the reader. This introduction should set out and describe the tasks that the writers have undertaken and may also give the number of hours or days allotted to their completion.

After the introduction, the report continues with the body, and for the reasons given above, this section is generally divided into Completed Tasks and Uncompleted Tasks.

Completed Tasks

In this section, you should set out the completed tasks in a logical order and devote a brief passage to each in which you explain that it has been completed, and give the date of completion.

Uncompleted Tasks

As you did for the completed tasks, you should give the uncompleted tasks in a logical order, and if any task remains uncompleted and is behind schedule, you must tell the reader what actions you are taking to bring the task up to date and what date you expect to have the task completed by.

Conclusion

At the end of the report comes the final section: the conclusion. The conclusion of a progress report differs from that of other reports, however, because it isn't a mere restatement of what has come before. It is premised on the earlier sections, but its purpose is to state whether the project is on schedule or not and, if not, why not. Furthermore, if the project is behind schedule, this is the place to state when it will be completed. It is helpful to readers for you to give an estimated completion date for the project even if it is on time so that they don't have to look this up or figure it out for themselves.

The conclusion is also the place to tell the reader whether any modifications to the project the project are necessary and, if any are, what they are and why they must be made. Finally, the conclusion is the place where you supply your name and contact information so that anyone with questions can contact you.

Additions

You may wish to put a brief abstract at the beginning of the report stating very briefly what the project is and whether it's on time or, if not, when it will be done. This information can usually be communicated in a couple of sentences and is very useful either to the reader who hasn't the time to read the entire report but wants the gist of it, or for readers who want a couple of sentences to help focus their minds and help them understand what follows.

A Gantt chart or a timeline may be helpful as well. If you must use a detailed Gantt chart, then append it at the end of the report. If you can make a more general one that takes up only a third or quarter of a page, then you can place it as a figure in the text of the report; a more general Gantt chart such as this may, in fact, be more useful to the reader than a highly detailed one which is, after all, a tool for those actually working on the project.

A schedule may be useful as well: this will constitute a table of tasks you must do and the dates by which you must do them. Ask yourself, however, whether you need both a schedule and a Gantt chart: the answer is almost certainly no.

PROGRESS REPORT TEMPLATE

[Memo Heading or Coversheet]

OVERVIEW IF NECESSARY

INTRODUCTION

Problem and why it is important

Description of project

List of tasks involved in completing project

- Statement of what is involved in performing the required tasks
- Time allotted or dates for completion
- Names of those doing the tasks (if appropriate)

Very brief summary of progress to date

Forecast (for longer reports)

TASKS COMPLETED

- Task A
- Task B
- Task C

TASKS REMAINING

- Task D
- Task E
- Task F

CONCLUSION

Prediction of when project will be done

Explanation of any changes that need to be made to the project

Contact information

OTHER MATERIAL THAT MAY BE HELPFUL TO THE READER

Abstract stating very briefly what the project is and whether it's on time or, if not, when it will be done (at beginning)

Gantt Chart (attached or inserted as figure)

Schedule of work

Project budget or cost

EXAMPLE OF PROGRESS REPORT

MEMORANDUM

From: Robert Wilkins, Electrical Engineer
Muncie Small Motor Plant

To: Raquel O'Malley
Vice-President of Operations
Trippett Motors, Inc.

Subject: Replacement of motors on manufacturing line at Muncie plant

Date: 28 August 2006

Introduction

The electric motors that drive assembly lines 1, 2 and 7 of our small motor plant in Muncie, Indiana have reached the end of their useful life, and they must be replaced. Those lines are involved in the production of carburetors for the small gasoline engines we supply to various manufacturers of lawn mowers, chainsaws and golf carts. On 20 July 2006 Mr. Delwood Parker, Vice-President of our Small Motor Division, assigned me the task getting the electric motors in those lines replaced and having the lines running with the new motors by 15 October 2006. The budget for the replacement is $12,000, which is to include the cost of the new motors, the cost of any re-wiring that may necessary if the electrical load of the new motors is greater than the old, the cost of removing and disposing of the old motors, and the cost of down-time while the three lines are stopped for installation of the new motors.

Project Description

This project involves eight main tasks:

1. *Determining what horsepower is needed to drive the three assembly lines.* This task involves calculating the power needed to move the volume of parts we are currently producing and those we expect to produce in the future. (6 hours)

2. *Determining whether new electrical lines for the replacement motors are needed.* This task involves a review of the wattage and amperage needed to power electric motors of the horsepower required. (5 hours)

3. *Researching what motors meet the requirements that we have.* This involves browsing industry catalogues, searching the Internet and telephoning company representatives. (8 hours)

4. *Getting an acceptable price for the motors.* We expect to negotiate several times with different company representatives. (8 hours)

5. *Purchasing the motors.* (3 hours)

6. *Installation of new electrical lines if necessary.* This involves discussions with our in-house electricians and having them install the lines. (7 hours)

7. *Installation of the new motors.* This will involve installation by our electricians and the shutting down of the production lines involved. (8 hours)

8. *Disposal of the old motors.* This task involves locating a place for the ecologically sound disposal of the motors that have been replaced, and then delivering them there. (2 hours)

Summary of Status

The project is going well so far and should be completed on 15 September 2006, a month ahead of schedule and $1,000 under budget.

Work Completed

The tasks below have already been completed.

Determining what horsepower is needed to drive the three assembly lines.

Assembly lines 1 and 2 are fitted with 2 hp electric motors manufactured by Bothwell, Inc. They were used without serious difficulty for the last eight years. Line 7 is fitted with a 1.5 hp motor built by Henderson, Inc.; investigation shows, however, that Line 7 is under the same load as lines 1 and 2 and that a 2 hp motor would therefore be a better choice and would likely last longer. Accordingly, I have determined that three 2 hp electric motors are required.

Task Completed: 23 July 2006.

Determining whether new electrical lines will be needed for the replacement motors.

No new wiring is needed for the installation of three new 2 hp electric motors. The existing motors draw between 22 and 25.6 amps under full load. The motors I have considered as replacements are of newer design and draw no more than 21.8 amps under full load.

Task Completed: 30 July 2006.

Researching what motors available to us meet the requirements that we have.

Three manufacturers make motors that will not require us to re-wire the lines; that is, they will use an acceptable amount of power. These motors all have acceptable warranties and an acceptable projected working life of between 12 and 15 years. They are of roughly comparable price. The manufacturers are Bothwell, Henderson and Tokagawa.

Task Completed: 5 August 2006.

Getting an acceptable price for the motors.

I contacted representatives of the three manufacturers and solicited prices from them. For the three 2 hp motors delivered to the Muncie plant the costs are

- Bothwell: $3,012
- Henderson: $2,899
- Tokagawa: $4,210

Although the Henderson price is slightly lower than the Bothwell, conversations with the Bothwell representatives have convinced me that we can get better support from them if we need it, and their warranty is 5 years long. Henderson and Tokagawa give a 3-year warranty. We can afford the Tokagawa motors in spite of their price, but their performance appears to be no better than that of Bothwell and Henderson.

Task Completed: 8 August 2006.

Purchasing the motors.

I purchased three Bothwell motors at the cost of $3,012 for delivery on 18 August 2006.

Task Completed: 10 August 2006.

Work Not Yet Completed

The tasks below have not yet been completed.

Installation of the new motors.

I have met with our plant electrician, Mr. Dale Wilson, and he has scheduled the installation of the new motors for 12 September 2006. He expects no difficulties in installing the motors; each line will have to be shut down for approximately 3 hours. The cost of this down-time has been calculated at $8,000.

Disposing of the old motors.

I contacted the Muncie Technical Institute and offered to donate the old motors to them so they may use them in their electricians' training program. They have accepted and will send a van to pick them up on 15 September 2006. Accordingly, there will be no charge for disposal of the motors.

Conclusion

The project is going well and is currently ahead of schedule. No changes in the project have been necessary. If things continue as planned, the work will be completed a month early, on 15 September 2006, at a cost of about $11,000, which is approximately $1,000 under budget.

If you have any questions, please contact me at: robwilkins8@munciemotors .com.

5

LABORATORY OR TEST REPORTS

Purpose of a Laboratory Report

The purposes of a laboratory (test) report are to provide (1) a clear explanation of the goals of an experiment or experiments; (2) all of the necessary details about the experiment, including the materials used, the exact procedure, and any errors that might have occurred; and (3) a detailed explanation assessing and relating the compiled results and data from the test in order to provide conclusions and/or recommendations based on the test results. The document needs to be extremely clear about every detail so that managers can make decisions on the basis of the results, conclusions and recommendations and so that other researchers can duplicate and further develop tests.

Preparation for Writing a Laboratory Report

Prior to writing a test report, you should develop a clear description of the test goals and create a detailed account of the experiment. Also, you should document details about the materials and equipment, the exact test procedure, observations, and experimental results. Also, you may choose to include theory related to the experiment. The theory should be clearly referenced (see **Appendix A** for guidelines on the use of sources).

Front Matter and Overview for a Laboratory Report

The laboratory report typically has a memo heading followed by a foreword and summary (see Chapter 1) that present the main findings for a particular audience, such as managers or executives. Alternatively, it may begin with just a cover page and table of contents [See **Appendix F** for descriptions of these elements.]

Main Body Sections

The test report contains a simple introduction that describes the specific goals or objectives of the experiment. The simple introduction may also include relevant background information about the experiment, such as previously conducted research or tests, and a 'forecast' of the organization of the remaining sections of the report, especially if the report is longer than 2-3 pages. A forecast is a statement of topics in the order that they appear in the body section. The forecast statement offers the reader a preview of the topics.

The body sections and sub-sections of a test report usually include these elements:

Background (sometimes combined with the simple introduction)

The background section offers the reader general information about previous work that may have been completed or related materials. It helps the reader to place the current work in context.

Theory (sometimes combined with a background section)

The theory section includes principles or mathematical equations related to the test or experiment.

Procedures

Materials and Equipment [a list of physical items used in the test, including equipment]

Procedural Steps [a list of detailed steps conducted during the test]

Data Collection [a detailed account of the methods used for collecting data]

Procedural Errors [a description of errors which have occurred during the test procedure]

Tips and Warnings [suggestions for conducting future tests]

Results and Analysis

The Results and Analysis section contains a full account of the data and a comprehensive discussion regarding the interpretation and implications of the data.

Closing Section

A conclusion section summarizes the main finding(s) of the test and reiterates the implications of the findings. This section may also include a recommendation for future related work.

LABORATORY REPORT (TEST REPORT) TEMPLATE

INTRODUCTION

Objective of the test

Brief background, if applicable

Forecast

BACKGROUND

General information about previous work or related materials that helps the reader to place the current work in context

THEORY

Principles or mathematical equations related to the test or experiment

PROCEDURES

- Materials and equipment
- Procedural steps
- Data collection methods
- Procedural errors
- Tips and warnings (if applicable)

RESULTS and ANALYSIS

A full account of the data and a comprehensive discussion regarding the interpretation and implications of the data

CONCLUSION

A summary of the main finding(s) of the test and reiteration of the implications of the findings

Discussion of future work, when necessary

ATTACHMENTS

[Raw data, graphs, charts, tables, related articles]

EXAMPLE OF A LABORATORY REPORT

To: Mr. Daniel West, Design Team Manager

From: Shari Hannapel, Design Engineer

Date: November 28, 2007

Subject: Two-Dimensional Foil Lift and Pressure Survey

Attach.: Appendix A: Water Tunnel Velocity Curve

Appendix B: C_P and $(V/V_0)^2$ Data

Appendix C: Plots of C_L Numerical Integration

Appendix D: Error Calculations for V/V_0

Foreword

The concepts of lift and drag were studied for a NACA-0015 foil for the Arsenal Ship Project. The lab results from this test will determine whether our department can approve the foil for production. The velocity profile on the surface of the foil was measured to determine the lift coefficient of the foil. The drag on the foil was determined from the wake velocity by application of conservation of momentum and mass. The purpose of this report is to present the test procedure, analysis, and recommendations.

Summary

A NACA 0015 foil was studied in a water channel. The foil was positioned in water flow at two different velocities and at three different angles of attack. The pressure along the surface of the foil was studied to determine the lift coefficient and to compare the experimentally obtained value with theoretical values.

The experimental values for the lift coefficient are close to the theoretical values, but not within error for positive angles of attack. The experimental values are also not within error of the theoretical values for negative angles of attack.

Therefore, without further testing, our team cannot recommend that the current foil design continue into the production process.

1 Introduction

The objectives of this laboratory were (1) to determine the lift coefficient and drag on a foil in uniform flow and (2) to compare these values to the theoretical values.

2 Background and Theory

This section presents the theory relevant to the pressure on immersed bodies and the determination of velocities from pressure measurements that serves as the basis for this laboratory and the data analysis presented in this report. Linear thin foil theory is also discussed.

2.1 Pressure on Immersed Bodies

The pressure and velocity of horizontal, steady, inviscid flow vary according to Bernoulli's equation:

$$p + \frac{1}{2}\rho V^2 = \text{constant} \qquad [1]$$

where p is the pressure, ρ is the fluid density, and V is the fluid velocity, along a streamline. When the flow reaches a fixed body, some fluid will flow above the body, and some fluid will flow below the body. One streamline will continue directly toward the body, and the fluid velocity at the body will be zero. The point where the velocity is zero is a stagnation point. Then, according to Equation [1], the pressure at the stagnation point, p, will be . . . [Paragraph should continue with further details. Additional equations should be inserted as above at the appropriate points.]

3 Test Procedures

The section below treats the apparatus and procedures used in this experiment.

3.1 Apparatus

The experiment was performed in the water tunnel (actually a channel) in the Marine Hydrodynamics Laboratory. The water tunnel is constructed of 1/4-inch mild steel with a steel frame and holds approximately 2400 gallons of water.

Flow was generated using 0.787 m diameter, four-bladed bronze impeller. The impeller was driven by a 20-hp three-phase induction motor. The motor speed was controllable and had a maximum of 1750 rpm. The motor speed was measured using a magnetic pickup with a 60-tooth wheel, and the speed was displayed on a digital counter. The water speed versus motor speed curve is available in Appendix A . . . **[Paragraph should continue with further details.]**

3.2 Experimental Procedure

The water tunnel motor was set to a constant speed. The foil was rotated until the stagnation pressure reached a maximum. The speed and pressure along the foil were recorded. At the same motor speed, the foil was rotated to 6° angle of attack, and the pressure was recorded along the foil. The foil was rotated to an angle of attack of -6°, and the pressure along the foil was recorded again. The procedure was repeated for a different speed.

Measurement of the wake could not be performed because of equipment malfunction . . . **[Paragraph should continue with further details.]**

4 Results and Analysis

The data discussed in this analysis are from the lab session conducted on December 14, 2007.

4.1 Foil Velocity

The zero angle of attack was determined by means of adjustment of the foil until the stagnation pressure reached a maximum. When the pressure tap at the leading edge is directly aligned with the flow, the pressure tap measures the stagnation pressure. The stagnation pressure is the largest measurable pressure in the flow, because of Bernoulli's equation as given in Eq. [1] . . . **[Paragraph should continue with further details.]**

The results for C_P and $(V/V_0)^2$ are tabulated in Appendix B.

From the results for $(V/V_0)^2$, V/V_0 can be plotted along the length of the foil. Plots of V/V_0 are included in Figures 1 to 6 for varying angles of attack and varying speed. Generally, three angles (-6°, 0°, and 6°) and two speeds were tested . . . **[Paragraph should continue with further details.]**

(graphic for Figure 1)

Figure 1: V/Vo along the length of the foil at an angle of attack of negtive six degrees and at high tunnel speed (715 rpm motor speed).

A theoretical foil of zero thickness would have no form drag. Form drag is related to the change in pressure, and by Bernoulli's equation, the change in pressure is related to a change in velocity. . . . **[Paragraph should continue with further details.]**

The results presented in Figures 2 and 3 show a comparison between the experimental results and the theoretical data for a NACA 0015 foil. V/Vo is zero at the leading edge of the foil, which . . . **[Paragraph should continue with further details. Figures 2 and 3 would be inserted at appropriate points.]**

The experimental data agrees with the theoretical data. For both the lower and higher speed cases, the experimental values for V/Vo follow a similar trend . . . **[Paragraph should continue with further details.]**

To calculate the lift coefficient for nonzero angles of attack, D was integrated along the length of the foil (on one side only) according . . . **[Paragraph should continue with further details.]**

5 Error Analysis

The section below treats the types of errors that affected experimental results. These include scale error, quantizing error, systematic errors, and errors in V/Vo and C_L..

5.1 Error in Pressure Measurements

Two types of error are quantitatively included in the error analysis for the stagnation pressure measurements and four of the foil surface pressure measurements: scale error and quantizing error. Scale error is calculated . . . **[Paragraph should continue with further details.]**

Systematic Errors

Several systematic errors that may affect the results of this experiment exist. First, there are some problems with the construction of the water channel. The water channel may not produce good laminar flow because of its design. The airfoil is not perfectly made . . . **[Paragraph should continue with further details.]**

Second, pressure measurements were recorded at discrete points along the foil. The true lift coefficient is calculated by approximating an integral along the surface, . . . **[Paragraph should continue with further details.]**

Finally, there were problems with a part of the apparatus. The angular displacement transducer was . . . **[Paragraph should continue with further details.]**

5.3 Error in V/V_0

The ratio of the velocity along the surface of the foil, V, to the free stream velocity, V_0, can be determined from . . . **[Paragraph should continue with further details.]**

5.4 Error in C_L

The lift is calculated theoretically according to Eq. [9], which involves an integral of the pressure along the entire surface of the foil. In this experiment, the pressure was measured . . . **[Paragraph should continue with further details.]**

6 Conclusions

A NACA 0015 foil was studied in a water channel. The foil was positioned in water flow at two different velocities and at three different angles of attack. The pressure along the surface of the foil was studied to determine the lift coefficient and compare the experimentally obtained value with theoretical values.

The experimental values for the lift coefficient are close to the theoretical values, but not within error, for positive angles of attack. The experimental values are also not within error of the theoretical values for negative angles of attack.

On the basis of the information from this test, our team cannot recommend that the current foil design continue into the production process without further testing.

6

DESIGN REPORTS

Purpose of a Design Report

The main purpose of a design report is to provide the details or a description of a device or process. The purpose may also include a discussion of the logic of the design or the reasoning behind its development. The design report is often a persuasive document because the writer must convince the audience that the design of the device or process meets important criteria that have been established during the development of the design.

Preparation for Writing a Design Report

During the development of the design, the designers have established design criteria or design standards that need to be met. Some general examples of criteria may include feasibility of the design, manufacturability, functionality, usability and cost. Some 'design-specific' criteria may include size, weight, speed, color, durability, efficiency, environmental sustainability, and marketability. Before writing a design report, you should define the criteria so that you can effectively explain how you arrived at your design and show that it meets the criteria.

Front Matter and Opening Sections

A cover page, table of contents and an executive summary precede a design report. The executive summary provides a synopsis of the main findings for a particular audience, such as managers or executives. It should also provide a clear explanation of the need for the design, a brief description of the design process, an overview of the entire design and an explanation of the main features and benefits of the design.

Main Body Sections

The design report begins with a simple introduction to explain the main purpose of the design and to describe its major components, features and benefits. Although the contents of a simple introduction may seem to overlap with the contents of the executive summary, the two sections serve different purposes: the recipient of your report can read the executive summary without having to read the design report or, conversely, may read the design report without reading the executive summary. Without the simple introduction, therefore, the design report would be incomplete. The introduction also includes a brief explanation of the structure or 'forecast' of the body of the paper. A forecast is a statement of topics in the order that they appear in the body section. The forecast statement offers the reader a preview of the topics.

Following the introduction, the major sub-sections within the body of a design document are the following:

1. Relevant background information (if necessary) that should orient the reader and place the design issues into context
2. Definition of all relevant design objectives and criteria
3. Design description, including
   a. detailed features and benefits
   b. explanations of how your design meets the criteria
   c. major design decisions and tradeoffs (or methodology)
   d. alternative designs that you considered and the reasons you rejected them

    e. potential drawbacks to the design (including a refutation to these drawbacks)

  4. Budget (if necessary)

Closing Section

A conclusion section for a design document explains the main features and benefits of the design and reiterates how the design meets the established design criteria.

References

All research reports will have a list of sources and will use some standard system for citing and documenting those sources. For further discussion of techniques for handling and citing sources, see the section called 'Making Effective Use of Source Material' in **Appendix A.**

Types of Attachments

Research reports often contain appendices; typical content for these includes some of the items listed below:

Detailed results of experiments, surveys, etc. discussed in the body

Detailed specifications for devices discussed in the body

Additional tables, figures, and graphics to supplement those presented in the body

Glossaries or extended definitions of unfamiliar terms and concepts

DESIGN REPORT TEMPLATE

COVER PAGE

Descriptive title

Name(s) of the designers

Date

Name(s) of the persons that the report is being submitted to (optional)

EXECUTIVE SUMMARY

A clear explanation of the need for the design

A brief description of the design process

An overview of the entire design

An explanation of the main features and benefits of the design

TABLE OF CONTENTS

Detailed listing of topics and subtopics with section numbers and page numbers

SIMPLE INTRODUCTION

Need for the design

Introduction to the design

A forecast of the rest of the report

BACKGROUND INFORMATION

Relevant background information, as necessary

DESIGN CRITERIA AND SOLUTION DESCRIPTION

Definition of design criteria, standards or goals

Design description, including

a. detailed features and benefits

b. explanations of how your design meets the criteria

c. major design decisions and tradeoffs

d. alternative designs that you considered and the reason you rejected them

e. potential drawbacks to the design (including a refutation of these drawbacks)

Budget (as necessary).

CONCLUSION

Main features and benefits of the design and reiteration of how the design meets the established design criteria

REFERENCES

A detailed list of sources cited in the report or consulted by the writers, keyed to the citations in the text

APPENDICES (if needed)

EXAMPLE OF DESIGN REPORT

Gold Lake Shoreline Redesign

Date Submitted:

15 December 2007

Prepared by:

Armaan Pirzada

Eleanor Wang

Karina Bary

Environmental Assessment Engineering Company

2121 Main Street

West City, Kansas

Prepared for:

Ms. Sandra Beech, Superintendent

North County Parks Department

Table of Contents

Executive Summary

Over the past thirty years, the ecosystem around the North County access ramp to Gold Lake has deteriorated. This deterioration has harmed the fish and wildlife that inhabit the area, and the North County Parks Department asked our team to develop a design solution that would improve the habitat at the site while preserving its existing uses.

Therefore, our team visited the site and developed a plan to reduce the overall environmental impact on the shoreline. Our goal was also to develop a design that would be aesthetically pleasing and would also provide potential recreational facilities. Also, we wanted our design to meet various criteria, including safety for the users, a maximum redesign cost of $100,000, and low maintenance requirements. We researched several options for incorporating a rocky beach, a breakwater and a boardwalk at the site. We also evaluated various aquatic enhancements and docks. Finally, we completed a financial analysis to determine the best design options for the site.

Our final shoreline design consists of three main components: a sloping rocky beach, an offshore breakwater, and a boardwalk. The total cost of these components is estimated to be $55,000.

In the proposed design, our team recommends that the existing metal breakwall remain intact because the cost of removal exceeds our budget restrictions. We recommend implementing a sloping rocky pile in front of the wall that will span the length of the shoreline (approximately 250 feet). This pile will be 7 feet high from the base of the sea wall and will slope into the water for about 70 feet. Two different types of rocks will make up the pile: the first, concrete rip-rap, will make up half the tonnage. Rip-rap generally occurs in large random-sized pieces. The second half of the tonnage will be made up of 3- to 9-inch stones. Different sizes of stone are desirable because they create a suitable breeding environment for fish and reduce wave reflection.

Our team also recommends the integration of an offshore breakwater in the form of fence lattice created from recycled plastic materials and connected by recycled piping. The lattice will be secured to the sea floor with rip-rap anchors in 12 different locations. The material for the lattice is low-cost and easy to maintain or repair.

In the final design, we also recommend building a boardwalk to camouflage the rusting breakwall that currently exists at the site. The boardwalk will be constructed of recycled plastic floorboards. The boardwalk will be 6 feet wide and 250 feet long. The boardwalk will also include 4 park benches and a 4-foot high railing along one side.

This report includes our final design, the criteria underlying it, and a budget for the completed work.

1.0 Introduction

The North County access ramp to Gold Lake was built in 1978 and has been in continuous use since then. A number of problems have arisen during the thirty years it has been in service. First, the original designers of the access ramp did not consider its impact on the area's ecosystem when they designed it, and the ecosystem has been damaged as a result. . . . **[Paragraph continues with further details.]**

Our proposed solution to these problems consists of three distinct elements: a sloping beach, an offshore breakwater, and a boardwalk, a . . . **[Paragraph continues with further details.]**

The report will begin by presenting the design objectives and criteria. Then, it will present the design specifications. . . . **[Paragraph continues with further details.]**

2.0 Design Objectives and Criteria

We established specific design goals and criteria for the site redesign with the client.

Our overall goals include redesigning the shorefront park to

1. restore the ecosystem to its original condition
2. preserve the current uses of the site
3. improve the habitat.

In response to these goals, we established the following design criteria. The design must

1. attain sustainable development
2. keep cost of the redesign under $100,000
3. ensure safety for users
4. preserve the current uses of the site
5. provide for functional requirements
6. ensure ease of maintenance of the shoreline redesign

2.1 Definitions

Key terms in these criteria are defined below:

"Sustainable Development"

Sustainable development is achieved when the patterns and intensities of resources used . . . **[Paragraph continues with further details.]**

"Cost of the Redesign"

The state government has allocated a maximum of $100,000 for this project . . . **[Paragraph continues with further details.]**

"Safety for Users"

Because boaters use the site extensively, they must find it easy to maneuver around the site, even at night . . . **[Paragraph continues with further details.]**

3.0 Design Specifications

The final redesign [Figure 1] of the Gold Lake access site consists of three main components:

1. a sloping beach made up of rip-rap and smaller stones
2. an offshore breakwater
3. a boardwalk

Each component is integrated with the design of the other components. Figure 1 shows the components of the final design as they will appear when the project is complete.

(add graphic of final design with labels, as appropriate)

Figure 1: Final design plan, including sloping beach, offshore breakwater, and boardwalk.

3.1 Rocky Beach

A total of 1,200 tons of rock will be used in the construction of the beach. This total is split equally between rip-rap and stones ranging in size from 3″ to 9″. The rocky structure begins at the existing seawall and slopes out to a distance of 70 feet. The rocks are positioned up to 7 feet high from the bottom of the lake. The main purpose . . . **[Paragraph continues with further details.]**

The positioning of large pieces of rip-rap side by side, interspersed with stones, creates small cavities. This arrangement produces a good habitat for small marine creatures and young fish . . . **[Paragraph continues with further details.]**

3.2 Boardwalk and Breakwater

The boardwalk will run the entire length of the shoreline: 150 feet along the eastern shore and 100 feet along the southern shore. Three main factors determine the plan for the construction of the boardwalk: safety, appearance, and durability. To address safety, a railing runs along the entire length of the boardwalk . . . **[Paragraph continues with further details.]**

3.3 Alternative Designs

We considered suggesting a concrete breakwater instead of the recycled plank breakwater. The alternative design was not practical because it failed to meet the cost criterion . . . **[Paragraph continues with further details.]**

3.4 Potential Drawbacks

Our team has identified two potential drawbacks to our recommended design . . . **[Paragraph continues with further details.]**

4.0 Budget Overview

The state government has given the North County Parks Department a $100,000 grant for a redesign of the Gold Lake Shoreline Facility.

Our estimate for the proposed redesign project is provided in Table 1. This table shows that our proposed solution to the problems at the Gold Lake facility would remain within the budget set by North Township.

Item	Cost
Rocky beach	$19,400.00
Boardwalk and breakwater	$35,550.00
Total cost	**$54,950.00**

Table 1: General budget for the redesign of the Gold Lake Shoreline Facility

We calculated these costs on the basis of . . . [**Paragraph continues with further details.**]

5.0 Conclusion

Our final shoreline design consists of three components: a sloping rocky beach, an offshore breakwater, and a boardwalk. The total cost of these components is estimated to be $55,000.

The design meets the criteria listed in Section 2.0, which were established by the client and our project team. The shoreline redesign improves the natural ecosystem, is easy to maintain and is aesthetically pleasing. We have complied with all relevant safety standards and have carefully observed all state rules and regulations. The design preserves the recreational uses of the site. It is also the most economical design that meets all the other criteria.

References:

[**All necessary references to documents, interviews, e-mail and other sources of authority would be listed here.**]

Appendices

[**Any necessary appendices would follow after the final page.**]

7

RESEARCH REPORTS

Purpose of the Research Report

The purpose of the research report is to give a supervisor or other interested reader detailed, well-supported information that you have gathered from reliable, well-documented sources, digested thoroughly, and synthesized usefully. When you write a research report, you should keep three imperatives in mind:

1) The research report should not be a string of direct quotations or close paraphrases;

2) You must maintain a clear distinction between evaluative or interpretive language and objective reporting of other writers' ideas;

3) In most instances, the overall rhetorical approach should be informative rather than persuasive.

The research report often takes the form of a long formal report, as in the sample report given here. For a brief survey of features specific to the long formal report, see **Appendix F.** Research reports may also, if short, take the form of a memorandum; for an example of the short research report in memo form, see the third sample memo in Chapter 2.

Typically you would produce research reports in a wide range of situations (apart from a classroom setting wherein you're assigned the task of exploring a topic for pedagogical purposes): when your company is in the early stages of the design process for a new device, process, or product and the head of the project needs a survey of previous approaches to the same problem; when you need to know the current state of knowledge in a particular field; or when you are choosing among alternative ways of solving a problem and need to know the historical performance of one or more of the candidates, for example. In short, when your primary focus is work that others have already accomplished or are in the process of doing, rather than work that you have done or intend to do yourself, you're going to need to produce a research report.

Opening Sections

Abstract (optional)

Overview

Introduction (see Chapter 1, Section 1.3)

> The subject of the report

> The scope of the report

> Your research methods, procedures, and goals (if appropriate)

> A forecast of the report's contents

Main Body Sections

Introduction to research topic

> Background: what was the research question your sources investigated?

> Position: how does the research question fit into the field in general?

> History: how has the question developed? What have been the approaches to it?

Exploration of research topic (see below)

Common content and patterns for the main sections

Examinations of individual studies, experiments, developments, models of a device, etc. [ordered chronologically or according to their importance to the reader].

Examinations of individual aspects of a particular process, device, theory or model, etc. [typically arranged spatially (device), chronologically (process), (model), cause-effect, or general-to-specific—see **Appendix A** for discussion of modes of development].

Closing Section

A typical concluding section for a research report will include some or all the following:

Restatement of the goals of the report

Restatement of the research topic

Restatement of the most significant findings

Clarification of the relevance of the findings for the managerial concern or problem

Suggestions for the uses of the research findings
Suggestions for additional or related research

Note that not all research reports require formal conclusions.

References

All research reports will have a list of sources and will use some standard system for citing and documenting those sources. For further discussion of techniques for handling and citing sources, see the section called 'Making Effective Use of Source Material' in **Appendix A.**

Types of Attachments

Research reports often contain appendices; typical content for these includes some of the items listed below:

Detailed results of experiments, surveys, etc. discussed in the body

Detailed specifications for devices discussed in the body

Additional tables, figures, and graphics to supplement those presented in the body

Glossaries or extended definitions of unfamiliar terms and concepts

RESEARCH REPORT TEMPLATE

COVER PAGE

Descriptive title

Name(s) of the writer(s)

Affiliation(s) of the writers (optional)

Name(s) of the recipient(s)

Name of the organization for which the report was prepared (optional)

Date of submission or dates of preparation

OVERVIEW: EXECUTIVE SUMMARY

A statement of the report's purpose

A clear explanation of the research topic and scope of the research

An explanation of the research methods, if appropriate

An overview of the research findings

TABLE OF CONTENTS

Detailed listing of topics (corresponding to section headings) and page numbers

INTRODUCTION

Context of the research

Introduction of the specific research topic

Statement of the goals, methods, and scope (as appropriate)

Forecast of the report's contents

BACKGROUND

Relevant background information, in as much detail as necessary

DETAILED DISCUSSION OF RESEARCH FINDINGS

Presentation of research findings (various sorts)

Discussion, analysis, and evaluation of findings (optional)

CONCLUSION

Restatement of report goals

Restatement of research topic

Reiteration of major findings

Clarification of relevance to audience

REFERENCES

A detailed list of sources cited in the report or consulted by the writers, keyed to the citations in the text

APPENDICES (if needed)

EXAMPLE OF RESEARCH REPORT

TITLE PAGE (See Appendix F)

(new page)

Table of Contents

Executive Summary

The purpose of this report is to provide details on the design, current status, and potential applications of the WIMS micro gas chromatograph.

Gas chromatography, the chemical analysis of complex gas mixtures by devices that collect atmospheric samples and then separate, concentrate, and identify the constituent elements, is widely used for monitoring levels of pollution in urban settings and for various safety and security purposes in military and industrial applications.

Conventional gas chromatography, however, is a slow and expensive process, and standard chromatographs are so cumbersome that their use is largely confined to laboratory settings. A need for smaller, cheaper gas chromatographs that are highly portable, require minimal power and can give results in seconds rather than minutes has become evident.

Development of a micro gas chromatograph, however, has depended on the development of the technologies that permit miniaturization of such complex devices. On the basis of this recent expansion in the fields of micro-electromechanical systems (MEMS) and wireless integrated micro-systems, researchers at the University of Michigan's WIMS/ERC have developed a prototype of a micro gas chromatograph that permits fast, accurate detection of pollutants and toxic gases in a wide range of environments and situations. Because of its small size, low power consumption, increased operation speed and programmability, the WIMS device has expanded the practical applications of gas chromatography.

The WIMS micro gas chromatograph incorporates the functions of full-sized gas chromatographs into a device miniaturized via techniques of micro-electromechanics. Conventional gas chromatography is a multi-stage process. These processes are carried out in the miniaturized WIMS device through the use of microelectromechanical systems.

Each of the miniaturized components acts in sequence as a fluid pathway in the minute-long chromatography process. In the first step, the vacuum pump collects gas from the surroundings. In the second step, the preconcentrator focuses the gas flow into a chamber until the desired concentration for detection is achieved. In the third step, the collected gas flows through separation columns at rates determined by their relative masses. Finally, in the fourth step, the mixture is analyzed by the particle detector, which sends information about the sample through wireless encoding. All of these processes are controlled by a microprocessor, which operators can reprogram remotely and control wirelessly. This feature permits changes in detection settings.

As of 2005, the individual components of the micro gas chromatograph were undergoing prototype tests, and a "Generation 0.5" prototype of the final device has now been completed. Specific future project goals call for an even smaller device with minimal power requirements, a large detection span, and a further reduced operating time.

Remaining impediments to wide-scale implementation of the WIMS micro gas chromatograph include special fabrication requirements and the need for improvement of the micropump and refinements of the sensors to maximize their reliability and sensitivity.

1 Introduction

Gas chromatography, the chemical analysis of complex gas mixtures by devices that collect atmospheric samples and then separate, concentrate, and identify the sample's constituent elements, is an indispensable tool in many industrial settings. It is widely used as a means of determining safe atmospheric composition for workers in high-risk settings such as mining and chemical production and rescue operations and as a means of monitoring pollution in urban environments such as Los Angeles and New York.

In recent years, the technology has found new potential applications in national defense and homeland security. Every year, millions of tons of cargo enter the United States through seaports without scanning for harmful gases. Chromatography could show when containers entering U.S. ports may contain hazardous materials. Furthermore, as awareness of the potential danger of chemical attacks has increased since 9/11, interest in the possibility of networks of gas chromatographs that could be installed in major cities, public gathering places, and sensitive industrial locations has grown.

Conventional gas chromatography, however, is a slow and expensive process, and standard chromatographs are cumbersome. A typical device is approximately the size of a microwave oven, and the most sophisticated models are considerably larger. Analysis of a sample can take up to five minutes or even longer. The costs and power demands are both high. These features have generally confined the use of the device to laboratory facilities. A need for smaller, cheaper gas chromatographs that are highly portable, require minimal power and can give results in seconds rather than minutes has become evident.

Development of a micro gas chromatograph, however, has depended on the development of the technologies that permit miniaturization of such complex devices. The last decade has seen tremendous expansion in the fields of micro-electromechanical systems (MEMS) and wireless integrated micro-systems. Funded by a National Science Foundation grant, the WIMS research group at the University of Michigan, Ann Arbor, is now developing dozens of individual micro-device components.

On the basis of this recent work, WIMS researchers have developed a prototype of a micro gas chromatograph that combines the principles of conventional gas chromatography, in which gases are separated by their flow characteristics, and microelectronics – small scale, low-power, high-function devices. The WIMS micro gas chromatograph design will permit fast, accurate detection of pollutants and toxic gases in a wide range of environments and situations (Agah, et. al, 2006). By reducing size and power consumption while increasing operation speed and programmability, the WIMS team has expanded the practical applications of gas chromatography (Agah et al., 2006).

The following report will outline the development of the WIMS micro gas chromatograph and provide details of the design and function of the current prototype. It will conclude with a brief discussion of some impediments to the implementation of the device and the directions of the ongoing research.

2 Background

As the section above notes, the WIMS micro gas chromatograph incorporates the functions of full-sized gas chromatographs into a device miniaturized via techniques of micro-electromechanics. Thus an understanding of this device requires a preliminary understanding of both conventional gas chromatography and the features of micro-electromechanical systems.

2.1 Conventional Gas Chromatography

Conventional gas chromatography is a multi-stage process (collection, concentration, separation, and analysis) that occurs in four regions or subsystems of the device: the intake pump, the preconcentrator, the separation columns, and the particle detector.

2.1.1 First Stage: The Intake Pump

To analyze gases in the ambient atmosphere, the device must first take in a sample. It does this by means of a vacuum pump . . . [The paragraph would go on to explain how this part of the process works. It would

proceed through the remaining material outlined in the section introduction above, labeling the subsections according to the numbering above.]

2.2 Microelectromechanical Systems

The key characteristics of microelectromechanical systems are captured in the triad "miniaturization, multiplicity, and microelectronics" (Helvajian, 1999). Miniaturization results in devices that are small, light, and fast. These characteristics ensure versatility of use. *Multiplicity* refers to the low per-unit cost of mass production; semiconductors are fabricated in large batches, so initial research and development costs can be recouped in volume sales. Finally, the microelectronic design, which combines sensors and input-output logic, produces devices that have the capacity for "decision-making" and reprogrammability, a feature which again leads to versatility (Helvajian, 1999).

2.2.1 Miniaturization

Miniaturization is achieved through . . . [**The paragraph would go on to explain the terms and concepts introduced above. The rest of the section would proceed through the remaining material outlined in the section introduction above.]**

3 Design and Function of the Wims Micro Gas Chromatograph

The individual sub-components of the micro gas chromatograph have been in continuous development since 2001. The four major components of the device, shown in Figure 1, are the gas-collecting vacuum pump, the pre-concentrator, the separation columns, and the particle detector. The size of the device is demonstrated by the U.S. nickel pictured beside it.

Each of these components acts in turn as a fluid pathway in the minute-long chromatography process. In the first step, the vacuum pump collects gas from the surroundings. In the second step, the pre-concentrator focuses the gas flow into a chamber until the desired concentration for detection is achieved. In the third step, the collected gas flows through separation columns at rates determined by their relative masses. This stage of the process can be controlled through the alteration of temperature and pressure in the columns. Finally, in the fourth step, the mixture is analyzed by the particle detector, which sends information about the sample through wireless encoding.

Figure 1: Components of the
WIMS micro gas chromatograph
(Wise et al. 2006)

All of these processes are controlled by a microprocessor, which operators can reprogram remotely and control wirelessly. This feature permits
changes in detection settings (Gordenker, 2005). Each step and the relevant components of the device will be explained in further detail below.

3.1 The Vacuum Pump

The vacuum pump, pictured in Figure 2, is the component that takes in
and directs the sample. It consists of two pumping chambers stacked on
top of each other, sandwiching a shared pumping membrane (Kim et al.,
2001). This structure, according to Kim et al. (2001), enables it to "accu-
mulate small pressures at each stage and achieve high gas pressure while

Figure 2: The vacuum pump
(Kim et al., 2001)

accommodating weak forces in the micro-domain." . . . **[additional technical details.]**

In order to control the gas flow in different modes, the micropump opens and closes incorporated microvalves. The entire unit is encased in two silicon wafers that protect . . . **[additional technical details.]**

3.2 The Pre-Concentrator

The pre-concentrator is responsible for concentrating the gases sufficiently for analysis. As Figure 3 shows, it consists of a chamber connected by micromachined channels to the vacuum pump . . . **[Physical description of the component would follow here.]** The pre-concentrator is also connected to the separation columns. Microvalves prevent backflow of the gas sample . . . **[additional technical details.]**

The pre-concentrator operates by . . . **[additional technical details.]**

3.3 Separation Columns

The function of the separation columns is to separate gaseous mixtures in mere seconds with extremely high selectivity and sensitivity in parts-per-billion (Potkay et al, 2006). In the WIMS device, these columns are grooves etched into the silicon and thermally isolated from the rest of the device . . . **[additional technical details.]**

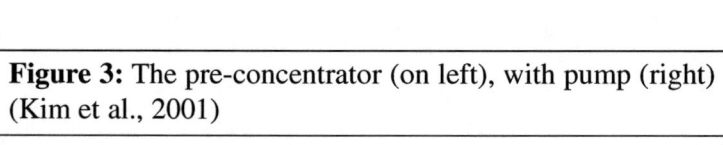

Figure 3: The pre-concentrator (on left), with pump (right) (Kim et al., 2001)

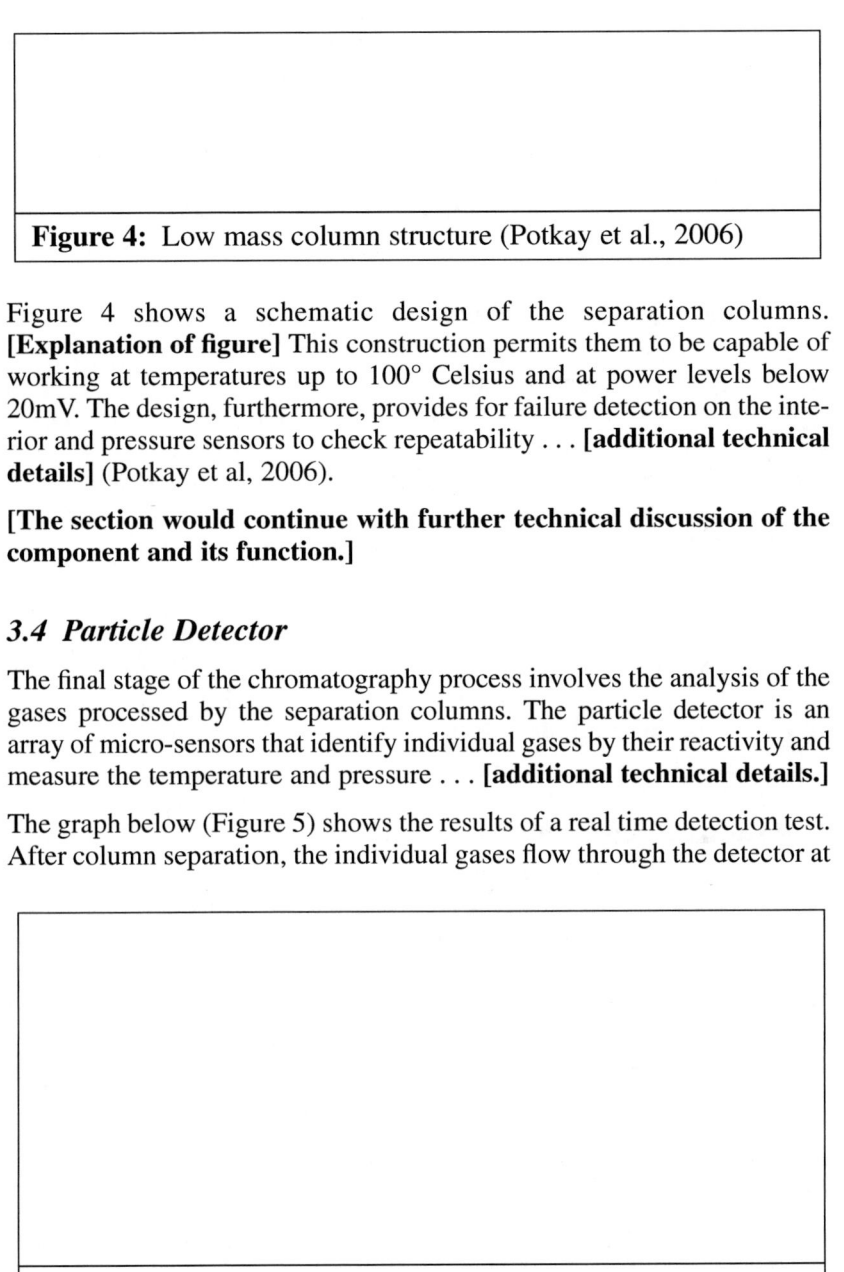

Figure 4: Low mass column structure (Potkay et al., 2006)

Figure 4 shows a schematic design of the separation columns. **[Explanation of figure]** This construction permits them to be capable of working at temperatures up to 100° Celsius and at power levels below 20mV. The design, furthermore, provides for failure detection on the interior and pressure sensors to check repeatability . . . **[additional technical details]** (Potkay et al, 2006).

[The section would continue with further technical discussion of the component and its function.]

3.4 Particle Detector

The final stage of the chromatography process involves the analysis of the gases processed by the separation columns. The particle detector is an array of micro-sensors that identify individual gases by their reactivity and measure the temperature and pressure . . . **[additional technical details.]**

The graph below (Figure 5) shows the results of a real time detection test. After column separation, the individual gases flow through the detector at

Figure 5: Rates of detection of varous gases (Gordenker, 2006)

different rates. The y-axis shows relative concentrations of ten discrete gases detected over the time axis . . . **[additional explanation of figure]** (Gordenker, 2006).

Through a series of alterations to the chemical composition of the electrode array in the sensors, the response of each electrode can be controlled and manipulated . . . **[additional technical details.]**

4 Current Status of the Design

As of 2005, the individual components of the micro gas chromatograph were undergoing prototype tests (Gordenker, 2005), and a "Generation 0.5" prototype of the final device has now been completed (Wise and Kim, 2007). Preliminary tests have shown impressive improvements in gas testing. While many early detection systems take as long as five minutes for sample analysis, the working prototype is capable of separating and detecting approximately twelve different gases in as little as fifty seconds (Zhong, 2006). Specific future project goals call for an even smaller device (one to two square centimeters in area), with minimal power requirements and therefore a smaller battery, a large detection span, and a further reduced operating time (Gordenker, 2005; Wise and Kim, 2007). . . . **[additional discussion.]**

5 Potential Applications

Current users of traditional gas chromatographs will benefit from the reduced cost and increased usability that miniaturization affords (see Table 1 for a comparison of costs and efficiency associated with the traditional and the miniaturized device). The device will enable a user to act as a fast, portable chemical lab, providing rapid testing and results for critical situations. One application for army personnel will be the testing for biological and chemical weapons in the field. Also, with miniaturization, rescue workers in hazardous situations could carry chromatographs embedded in their suits, an innovation that could lead to greater safety and better risk assessment . . . **[additional discussion.]**

Industrial applications will include the testing of gases in a fuel cycle of a fusion reactor . . . **[additional discussion.]**

Use of the devices to enhance national security will also be numerous. For example, individual micro gas chromatographs with wireless inter-

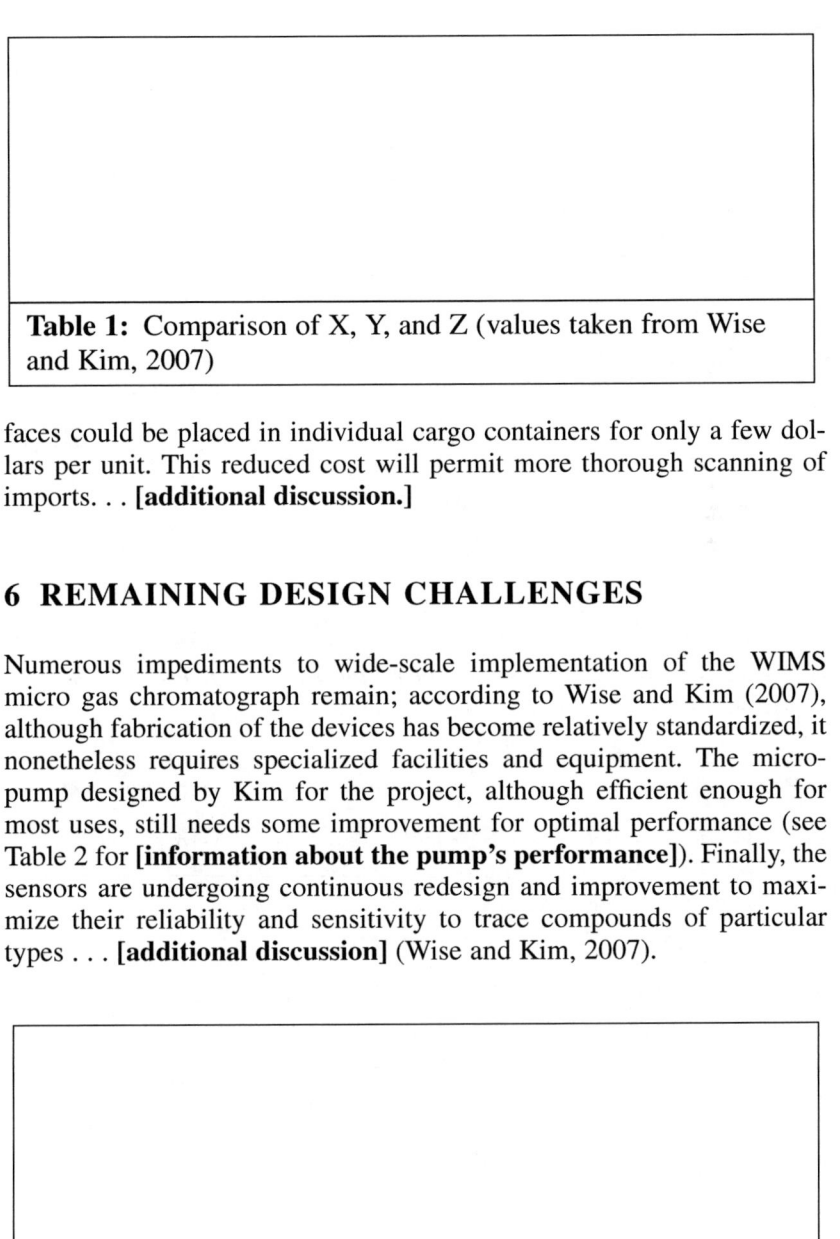

Table 1: Comparison of X, Y, and Z (values taken from Wise and Kim, 2007)

faces could be placed in individual cargo containers for only a few dollars per unit. This reduced cost will permit more thorough scanning of imports. . . [additional discussion.]

6 REMAINING DESIGN CHALLENGES

Numerous impediments to wide-scale implementation of the WIMS micro gas chromatograph remain; according to Wise and Kim (2007), although fabrication of the devices has become relatively standardized, it nonetheless requires specialized facilities and equipment. The micropump designed by Kim for the project, although efficient enough for most uses, still needs some improvement for optimal performance (see Table 2 for [information about the pump's performance]). Finally, the sensors are undergoing continuous redesign and improvement to maximize their reliability and sensitivity to trace compounds of particular types . . . [additional discussion] (Wise and Kim, 2007).

Table 2: Performance of Kim micropump (data from Kim 2006)

7 Conclusion

The WIMS Micro Gas Chromatograph project combines the potential of small, fast, low-power mechanical sensing and computing with an important environmental analysis tool: gas chromatography. Numerous advances by WIMS researchers have led to smaller, faster, more accurate detection, and these advances have great commercial prospects for industrial safety systems, pollution monitoring, and homeland security.

8 References

Agah, M., Lambertus, G. R., Sacks, R. D., & Wise, K. (2006). High-speed MEMS-based gas chromatography. *Journal of Microelectromechanical Systems, 15*(5), 1371-1378.

[ADDITIONAL REFERENCES HERE]

Gordenker, R. (2005). A MEMS micro gas chromatograph. *WIMS ERC Annual Report 2005*, 41.

Helvajian, H. (1999). *Microengineering aerospace systems*. Reston, VA: AIAA.

Kim, H. S., Preliminary study of micro-machined WIMS vacuum pump. Retrieved February 20, 2007, from http://www.wimserc.org

Kim, H. S., Bernal, L. P., Washabaugh, P. D. (2001). A micro-machined WIMS vacuum pump. Retrieved February 23, 2007, from http://www.wimserc.org

[ADDITIONAL REFERENCES HERE]

Wise, K. and H.S. Kim. (2007). *Current status of WIMS micro gas chromatograph*. WIMS/ERC internal document, not paginated.

Zhong, Q., Steinecker, W., & Jin, C. (2006). Portable mesoscale system for analysis of complex vapor mixtures. Retrieved February 23, 2007, from http://www.wimserc.org

8

ORAL REPORTS
(ORAL PRESENTATIONS)

Oral reports, or oral presentations, are an important and common feature of engineering and business. What follow are suggestions for choosing, organizing, preparing and presenting material orally.

Structure of Oral Reports

Oral reports are fundamentally different from written reports; keep in mind that when readers find something difficult in a report they can reread the difficult section as often as they need to. Furthermore, they can read a report when it suits them and take as much time over it as they want; consequently, a written report can make as many points as the author wishes. In oral reports, however, members of the audience must listen and try to take in as much as they can from a stream of speech, the speed and content of which they cannot control. You must organize your oral report in view of these realities.

The process of organizing an oral report involves three steps.

First, identify the purpose of your presentation. Like a written report, the oral report must state the topic, make its purpose clearly evident, and offer the audience a forecast, or layout of the report, in the introduction. You may want to follow your introduction with background information, if this information is necessary for your audience. Background should be presented only if it is relevant to the main points in the oral presentation and useful to your audience.

Next, consider what points you want the audience to recall after you've finished speaking. Choose the major points that you think are the most important and build your talk around them. Naturally, one of the basic ways to emphasize a point—to make it stand out—is to devote more text (or in the case of oral reports, more speaking time) to the point you want to bring out. Therefore, you should allocate more time—and thus more detail—to the points you want the audience to recall than to those other subjects that you might discuss only as background or introduction to your main points.

So, write out your major points, perhaps as claims or statements of fact, so that they are clear to you and well articulated. Then determine what support you need to establish these claims and what you need to say to develop these claims so they can be understood and remembered.

Finally, construct a meaningful conclusion that reiterates your main points. You can also thank your audience for listening and open the floor for questions from the audience, if appropriate.

Basic Organization of an Oral Report

Introduction
Background (if necessary)

Point 1
Relevant Support for Point 1

Point 2
Relevant Support for Point 2

Point 3
Relevant Support for Point 3
(Continue pattern with other points, if necessary)

Conclusion

Visuals for Oral Reports

Managers, clients and customers generally expect neat, professional graphics. Visuals, or presentation slides, can be useful to you and may effectively be required in your field. Furthermore, they can take the place of note cards as prompts for your talk, and if you practice with them you may find you can dispense with notes entirely.

Oral reports are too various in their forms, content, and style for us to give a set of rules for what slides you will need to include; some basic types, however, are all but required. Unless you have specific reason to think otherwise, assume that you will display slides of the following types:

- A title slide, which gives the title and date of the presentation and identifies the presenter(s) and their affiliation(s)
- An introductory slide, which provides some context for the material you are presenting
- An overview slide, which forecasts the content and organization of your presentation
- Content slides, as appropriate to the material
- A summary or conclusions slide, which reiterates the main points of your presentation
- A references slide, which provides bibliographical information for any sources you used in your presentation.

It's almost always a good idea to unify the slides with a consistent font, basic layout, and color scheme; a logo or design associated with your team or your company may appear on each as well. In designing the slides, keep the following guidelines for creating effective presentation visuals in mind (see also **Appendix C**):

- Put informative titles on all or most of your slides
- Use grammatically and logically parallel constructions in lists (see negative example on page 104)
- Maintain uniform font types and sizes on slides
- Use a font that is large enough for everyone in the audience to read

- Use images to illustrate textual descriptions
- Maximize contrast and sharpness of images
- Create simple visuals by using neutral backgrounds and clear text fonts
- Avoid putting too much text on one slide, but give sufficient information to make the slide meaningful to the audience
- Include citations or references on slides, when appropriate
- Avoid animation unless you use it to emphasize a point.

A slide with faulty parallelism might look like the one below. Notice that the fourth point, unlike the others, does not begin with a gerund. Not only is this deviation from the pattern disruptive, it prevents the audience from identifying a single consistent relationship among the points and between the points and the title. Revising the fourth point to read, e.g., 'making surgical incisions' would solve these problems simply. You should make seeking such solutions a habit.

> **Common Uses of Lasers**
>
> - Drilling Metal
> - Cutting Textiles
> - Measuring Distances
> - Doctors Use Them in Surgery
> - Reading DVDs

Lack of parallelism can occur when items in a list are not logically similar, too, and furthermore, lack of visual parallelism can result from needless variations in font style or size within a single section. While these may be less immediately obvious than lack of grammatical parallelism, they are also worth taking time to correct. The overall impression your presentation makes will be better if you provide your audience with the means to group related elements quickly and easily.

After your talk you may expect questions, and it's useful to antici-
pate these. As you prepare your talk, think what questions people might
ask and prepare to answer them. A good, solid answer to a question is a
good finish to a talk. If you plan to use presentation slides with your talk,
you might consider making a visual or two that would assist you in
answering a question you might be asked. Pulling one of these out of
your sleeve to answer a question can show the audience that you are
organized and prepared.

Speaking and Practice

Many people fear public speaking, and this is natural. The way to over-
come it is twofold: first, have something to say, and second, practice. If
you have nothing to say, then you have every reason to be uneasy. If
you've thought out what points you want to make and decided what you
need to say in order to make them convincing and memorable, though,
and if you've proportioned your talk properly to give the correct empha-
sis to the points, then you should not be afraid to speak for lack of anything to say.

Practice will remove most anxiety. By this, we mean that if you've
spoken a good deal in your career or in your life, you will most likely
be fairly comfortable with it. For this reason it's worthwhile to take
opportunities to speak in public. The more you speak in public, the more
self-assured you'll be, and the more your listeners will be comfortable
with you as well.

There's another aspect to practice, though: *to give a successful talk
you should practice that particular talk.* More precisely, you should
practice giving the talk out loud several times before you give it to your
audience. By "practice" we do not mean silently reading over a prepared
text or glancing over note cards again and again. We mean that you must
say the words out loud many times over a period of several days. A number of benefits arise from this practice: you'll learn your talk and con-
quer anxiety and, what's also very important, you'll get used to the
sound and feel of the words and sentences. As a result, your pronuncia-
tion will probably be clear, you won't stumble over unfamiliar sounds,

and your vocal inflections are likely to be natural. Speaking naturally is important, because it engenders confidence and disposes your audience to believe what you say.

If you are giving an oral report with multiple speakers, rehearse the presentation as a group by standing up and speaking. Make smooth transitions between speakers to produce a seamless presentation. Practice where each speaker will stand and how the computer will be used. These minor matters of choreography can make a presentation look polished– or not, if you fail to practice them and thus the transitions or coordination between speakers and slides are rough.

Here are some guidelines for delivering your presentation:

- Maintain eye contact with your audience
- Speak directly to the audience, using normal speech contours and rhythms
- Avoid reading word-for-word from the computer screen, visual, or note cards
- Avoid facing and reading from the computer screen
- Avoid standing directly in front of your visual while speaking
- Point specifically to your visual when you want to emphasize a point
- Maintain a professional tone and unhurried pace.

A

STRATEGIES AND GUIDELINES FOR CONTENT DEVELOPMENT AND ARGUMENTATION

Different types of content call for different arrangements of materials. For example, if your overall purpose is to recommend one of several competing designs, you clearly will need to do several different things: introduce the problem, provide any necessary background and history, enumerate and justify your selection criteria, describe the designs, compare the designs to each other, compare your choice to the criteria, present and justify your choice, and provide a brief conclusion. If your selection process involved testing the designs in some way, you will also need to explain what you did. Each of these subsections requires a different structure if it is going to do its job effectively.

Although the units we're talking about above are sections, for practical purposes you should be thinking about structuring paragraphs, and when you think about structuring paragraphs, you should think about two things: crafting a **topic sentence** that will control or direct the development and then building, sentence by sentence, the structure that the topic sentence sets up.

To see what we mean, consider the following three sentences:

a. *Three factors contributed to the failure of the electrical system.*

b. *The failure of the electrical system resulted from a series of preventable mistakes.*

c. *We propose to analyze the failure mechanism of the electrical system.*

Any of them could begin a paragraph in a report on a system failure, and all three lead the reader to understand that he or she is about to learn something regarding the causes of that failure. But would the rest of the sentences in the three paragraphs be identical, and would they come in identical order? If you really look at what the sentences say, you can see that they suggest different approaches and different content.

Before we look at the development process, though, you need to come to grips with that concept of *argumentation.* As an engineer, you'll learn to regard statements like the ones above as *claims,* and you'll need to form the habit of providing *support* for your claims. We'll talk further about that below. If you accept the proposition that an engineering document isn't often going to leave a statement like *Three factors contributed to the failure of the electrical system* or *The failure of the electrical system resulted from a series of preventable mistakes* to stand without any support, then you can begin thinking about what the support should be— i.e., how you should develop your paragraph.

The key is something most engineering students and professionals learned about in grammar school but forgot about as they moved into technical specialties: the topic sentence.

Topic Sentences and Paragraph Development

Well-crafted topic sentences generally contain a word or phrase that tips off the reader as to what to expect; inexperienced writers often include these words but fail to make use of them. What are the tip-offs in these example sentences? *Three factors. Resulted from. Analyze the mechanism.* These words tell the reader that Paragraph One will approach the subject by way of a list–in this case, of causative factors. Paragraph Two will approach it by way of identifying causes and their effects in the order that they occurred. Paragraph three will approach it by way of breaking down a process into its steps or subprocesses. As modes of development, these are called 'enumeration,' cause-effect,' and 'process analysis.' Other common modes are description (in which you lead readers visually through a physical object, a mechanism, or a process), narration (in which you tell what happened or what you did in chronological order), comparison/contrast (in which you can line up all the similarities between two things and then all the differences between them or move back and forth, noting similarities and differences as you look at the common features of the two things), and amplification (in which you give additional examples to illustrate something). If you go back to the example of the design report in the opening paragraph, you can see that some of the subsections enumerated there call for these different modes.

Once you have come to understand how the purpose of a paragraph dictates the structure that's most appropriate for developing it and how the topic sentence sets up the reader's expectations regarding the material that will follow and its arrangement, you can expand the principles to the section and then to the document as a whole. That is, major sections of a technical document typically have an opening paragraph that functions the way a topic sentence does. Each of the paragraphs within the section functions the way a sentence within the paragraph would; that is, it develops the 'topic paragraph' in the way the paragraph established. If you take the body of a well-written technical report, extract the topic sentences from each of the main paragraphs, and string them together, you should have a reasonably coherent paragraph that expresses the main content of the report—a skeletal one without the supporting evidence a technical document needs, certainly, but one that makes sense by itself. This is a good test; try it on one of your technical documents. If what you come up with doesn't make sense, your reader is going to have trouble following the report as a whole.

Any good college handbook will explain this material in greater detail and give you lots of helpful examples, but if you're willing to take the time to think about your purpose for writing a given paragraph, just these brief explanations should get you on the right track. Be aware, though, that sophisticated writers sometimes combine and alter the basic structures to suit their complex purposes, and thus sometimes you may not be able to identify a single mode in a particular paragraph. You should still try to organize your paragraphs and sections according to one of the patterns; that will make your reader's job (and yours) easier.

Developing Defensible Claims

Although many people assume that anything called an argument must involve heated disagreements between parties holding opposed opinions and thus that all claims must be statements of controversial positions, argumentation in engineering documents is most often a cool and impersonal process, and claims are most often nothing other than statements of the results of your investigation or research. In other words, argumentation is the process by which you demonstrate, by providing evidence and using logic, the plausibility or truth of conclusions that you have arrived at and now want the reader to accept.

Claims take different forms, depending on the type of document you're writing. The main claims of a proposal, for example, will almost certainly be

(1) that a suitable solution to a particular problem is 'X' and

(2) that you or your team is well prepared and well qualified to provide that solution.

These are conclusions that you will have reached through an analysis of the problem, an evaluation of what it will take to solve the problem, and an assessment of your own abilities to undertake the problem.

If you get the job, come up with a design, and submit a design report, you will have different claims in the latter; among them will almost certainly be (1) that certain criteria should be used to judge the success of a design solution and (2) that your design meets the criteria you have presented. If your task was to research a set of solutions that other engineers have proposed, your claims will probably include (1) above and that a particular design (or no design, or some combination of designs) best meets the criteria.

Within these sorts of documents, you will have many subsections that present sub-arguments; you may conclude that a particular type of sensor is best for a subsystem of the design, for example. Or you may conclude that the previous approaches to the design problem all overlooked a significant technical challenge. But the process remains the same: you present a conclusion that your work or your research led you to and enables you to support.

The most important quality of a good claim is that it is supportable (or 'defensible'), but several other qualities are also important. It must be narrow enough for you to support in the space that you have. For that reason, you should be very wary about including words like 'every,' 'all,' 'always,' 'none,' and 'never.' It must be as concrete as you can make it (for a discussion of concrete and abstract language, see Appendix B. It can't be self-evidently true, like a tautology or a statement of physical fact. It can't be a mere expression of personal taste. If you think about it, all these things make good sense. In the world of science and mathematics, defensibility depends on solid evidence and logical reasoning. You can and should assume your readers will look for them.

Selecting Good Evidence

As we say in Chapter 1, the best evidence in most engineering documents is quantitative; most conclusions you reach will rest on measurements, calculations, experimentation (yours or that of researchers whose results you find), and the like. You should assume that if your reader can reasonably expect numbers, you must, to argue effectively, provide them. Furthermore, you must usually interpret and analyze numerical data; long lists or tables of figures cannot, by themselves, make an effective argument.

Also as noted in Chapter 1, not every claim arises from operations with numerical data. Some arguments rest on inference or deduction. To make these sorts of arguments effectively, you should familiarize yourself with the basics of logical reasoning; any good textbook on logic can give you the information you need. Other arguments rest on qualitative data–observations of phenomena, for example. In any of these and other cases, two things remain constant:

- you must support your claims with evidence—the basis for your belief that the claims are reliable;
- you must make sure that the audience can see the connection between your evidence and the claim.

Many inexperienced writers grasp the first point but miss the second one. Thus they may make a claim like 'Sensor X is the best choice for this design' and follow it with a detailed set of the specifications for the sensor, forgetting that the audience can't necessarily conclude from those numbers that the claim is valid. Unless the reader knows *why* a sensor would need to have those particular specifications to function effectively in that application, the list of numbers doesn't help.

Many people who write about argumentation talk about 'data sets' or 'reasoning units'; by these terms they mean a group of sentences consisting of your claim, the evidence that supports it, and a statement that clarifies the relationship between that evidence and the claim. This formulation is a greatly simplified version of the so-called 'Toulmin model' of argumentation. It's a useful one to internalize, even though in actuality you may not always need that explicit link. Here's an example to help you understand the point:

> **Original:** *The most important properties of titanium are that it has high strength for its weight, it forms a protective TiO2 coating, and it can be machined with a high degree of precision.*

Is this an argument? Not really. It might be a claim, although it is rather broad and unspecific.

First revision:

Three properties of titanium make it suitable for use in biomedical applications. First, it has high strength for its weight; second, it forms a protective TiO2 coating, [and so on].

Is this now an argument? It's closer, because now, as you can see, it consists of two elements: a conclusion that the writer has reached (the claim) and a set of facts (evidence) that appear to support the claim.

But is it a good argument? Not very, yet. First, it doesn't use numbers where the reader might reasonably demand numbers—what is its strength-to-weight ratio, specifically? You'd want to supply a precise and accurate figure here.

Second (and harder for some people to see) it doesn't clarify why these properties have anything to do with suitability for use in biomedical applications. If the reader happens to know why, for example, a TiO coating is important in biomedical applications, he or she may sail right on by, possibly nodding agreement. But if the reader doesn't already know that, have you made an effective argument?

So the final version of this argument might look something like this:

Because metals used in biomedical applications must have high strength in relation to weight [precise numbers here would be helpful, too], *high resistance to corrosion* [a reference to position in the Galvanic series might be useful], *and good machinability to ensure very precise tolerances, titanium is a suitable candidate. Its strength-to-weight ratio is* [number]. *Its ability to form a protective TiO2 coating (passivation) gives it good corrosion resistance; passivated Ti is found at* [location] *in the Galvanic series. Its mechanical properties* [give numbers for ductility, strength, and hardness] *make it sufficiently machinable* [etc.]

You can see that the physical facts you list now actually support the claim in a way that the opening statement makes meaningful. That's what you want: a unit in which the data, or evidence, leads your reader to accept the claim that you make because the evidence is factual, precise, and relevant.

If you gave *only* the numbers for titanium's strength, ductility, and so forth, you'd have a different problem, though: however precise the data, unless the reader had a context in which the numbers were meaningful, they wouldn't persuade him or her that your conclusion was valid, would they? So what if titanium has a ductility of 17%, e.g..? Why is that important? How is it relevant? You can see that the connecting statement (called 'the warrant' in the Toulmin model) is crucial for making arguments work, even though it doesn't always have to be this explicit. In some settings, you can reasonably assume that your audience knows why certain data can support a given claim.

If you're sharp, you've probably noted that the first sentence in the revised argument is itself a claim that may require support. You have to work from assumptions. At some point, early on, you will probably need to lay out your working assumptions and let your audience accept or reject them. Of course, sometimes it's perfectly okay to proceed on the assumption that reasonable readers don't demand support for claims that conform to the working assumptions of your shared field of knowledge.

In short, however, you should always try to construct your arguments so that they are precise, well supported, and explicit as you can make them. Few readers will fault you for making things clearer than you *had* to.

The Importance of Credibility

The effectiveness of arguments in any scientific or engineering document rests largely on an essential quality that we'll call credibility. As any reader of science fiction knows, even writing that is full of quantitative data may not be reliable, and if you want your reader to accept what you say, you must demonstrate that you can be trusted. How do you do that? **First, you must recognize that readers judge your credibility to a certain extent on the basis of the appearance of your document**, and slipshod preparation or numerous careless errors will undermine the reader's faith that the writer has taken care to be accurate in the handling of the data in a report.

Second, you must sound like a member of the community you're writing for. That means that your diction must reflect the level of formality and precision that is customary in your field. **Finally, you must display authority.** You can acquire it by becoming a respected figure in your field. Until you reach this point, you rely on the work of those who have, i.e., you refer to material published or presented by authorities in your field. That involves making use of sources.

Making Effective Use of Source Material

If you are going to use material taken from sources, the first thing you should do is to establish a method for recording that material so that you can be sure, when you use it, that you have quoted or paraphrased it accurately and attributed it correctly. Don't think that you'll remember where an unattributed quotation comes from or that you'll remember exactly how a passage went if you just jot down a short version of it. Be rigorous in your bibliographical methods; misquotation or misattribution is sloppy at best and will get you into trouble at worst.

To use source material, you have to understand it. This may seem self-evident, but anyone who reads a lot of technical documents occasionally sees passages that have been inserted into texts with no apparent regard for meaning. If you do understand your source material, you can probably abridge, paraphrase, or summarize the relevant material, as your argument demands, and you can also tell your reader what its relevance to your argument is. These are processes, however, that require careful attention to sentence structure and to punctuation and graphic conventions. We'll talk more about those below.

Before we move on to a review of the rules that govern the way those parts of a technical report that uses source material should look, we need to go back to a basic principle: you should always credit your sources, accurately, thoroughly, and unambiguously. You can find extensive discussion of the reasons for doing so and vastly detailed rules for documentation in publications put out by most major American university libraries, on the website of the American Psychological Association, and elsewhere. If you are a student, your instructor will probably tell you what form of documentation to use. Technical professionals will need to select the form they use by looking at typical documents of the sort they are preparing (professional journals will supply you with their guidelines).

The two most common systems for citing material within a technical report are parenthetical documentation, which the APA website explains thoroughly, and numbered references, which the ASME site explains in equal detail. Software that keeps track of your sources, formats footnotes or endnotes as you type your report, and prepares a bibliography or list of references in the style you select is now widely available, also, and it makes the task of preparing a properly documented document much easier than it has been in the past. Some university libraries link their catalogues to a bibliography-creating software package; if you have access to a resource of this sort, you will find it very helpful in the mechanical aspects of this task.

We'll begin with a set of general guidelines for incorporating source materials into your texts and then look at the conventions for indicating the nature of your use of source materials.

1. **Never present a string of quotations in lieu of an argument. Source material is supposed to support your claims. It isn't a substitute for them.**

2. **Avoid lengthy quotations. Quote directly only when the way the source's author has said something is truly better than any way you could say it or when you want for some reason to call particular attention to the original material. Instead, paraphrase whenever you can do so without sacrificing accuracy and clarity. Often you can be clearer than the original, at least for your particular audience.**

3. **When you must quote directly, take only the essential parts of sentences from your source whenever you can. Quote full sentences only when you must for the sake of clarity.**

4. **Never just drop a quotation into your text. At a minimum, introduce full sentence-quotations with a brief statement like "A recent study gives statistics on the rate of failure: '[insert your quotation]'(reference)." Incorporate partial-sentence quotations into your own sentences. You'll have to pay attention and make sure that the resulting sentence still works grammatically.**

5. **If you quote directly, be absolutely accurate. Never alter quoted material without indicating that you've changed things (for rules on how to do that, see below).**

6. **When you use more than one source within a single paragraph, be very careful to distinguish them from each other and from your own material. It's best to do that in words rather than by relying on numbered references that the audience may not immediately notice consciously.**

Now we'll give some rules. Books that tell you how to do research papers or journal articles often have good collections of these guidelines. Remember that what you're doing is making sure that your reader can tell very clearly what material is yours and what is attributable to someone else. You're also making it easy for the reader to locate the original material in its complete form.

The rules for direct quotation of material are simple:

- **Enclose the quoted material in quotation marks if it is less than a sentence or two in length.**
- **Put the quoted material into a block (a paragraph indented at both right and left margins and single-spaced even when the document is otherwise double-spaced) if it constitutes several sentences or one long and complex sentence. Set the block off with an additional line above and below. Don't enclose it within quotation marks.**
- **Put a citation at the end of the block quotation; put a citation at the end of the sentence in which an incorporated quotation occurs. It will be outside the quotation marks but INSIDE the sentence; i.e., before the period at the end of the sentence.**

Follow the models closely until you have internalized the rules.

If you need only part of a sentence, the rules for abridging quotations are also simple:

- **Surround quoted material with quotation marks and indicate omissions with ellipsis dots (that is, three periods without intervening spaces, like this . . .). These are mandatory when you remove material from the middle of a quoted sentence;**

many writers leave them off when they pick up in the middle of a sentence or stop before the end; this is perfectly acceptable.

- Bracket anything you've had to supply to make the quotation make sense in the context of your sentence. Usually that means one of only a few things: a name to replace a 'he' or 'she' that wouldn't be clear, a noun or noun phrase after a 'this' that wouldn't refer to anything, sometimes an ending on a verb. This may be difficult to understand without specific examples; if you're in a technical communication class, your instructor will show you some.
- Put the citation at the end of the sentence that contains the quoted material.

Paraphrases and summaries of source material don't need special graphic indicators; the citations in the text tell the reader that they come from your sources.

Summarizing and paraphrasing source material, while essential skills, are things you learn by doing them, and success depends heavily on your grasp of sentence structure. Thus we direct you to Appendix B and any good textbook on English syntax if you need help in this area.

To close this section, we want to reiterate some basic precepts:

- Always let your purpose dictate the development of your units—paragraphs, sections, or entire reports.
- Always construct topic sentences and topic paragraphs carefully so that they guide your reader and forecast the content of the following unit.
- Construct your arguments carefully—develop claims that you can support and support them appropriately.
- Use source material appropriately, correctly, and honestly.

If you keep these precepts in mind as you're developing the content of your technical reports, you'll be on your way to producing effective technical communication.

B

STRATEGIES FOR EFFECTIVE USE OF LANGUAGE

Inexperienced technical writers often dismiss concerns with language as unimportant, on the grounds that engineers and scientists don't need fancy words or complicated sentences to communicate effectively. In one sense, they're correct. Using fancy words and complicated sentences generally conflicts with the primary goals of good technical communication: clarity, precision, and conciseness. But achieving those three goals often takes more attention, and sometimes requires more effort, than writing in a 'fancier' style. It takes a conscious application of principles. In this section, you will find a review of the most important ones.

You should know from the outset that you will **not** find a thorough treatment of grammar, syntax, or punctuation in this appendix. Many excellent books on the rules of Standard English are available, and everyone who writes as part of his or her job should have at least one of them on hand. Your instructor, if you are a student, can recommend some titles; your company library, if you are employed, will probably contain several as well. If you have never received formal instruction in grammar, sentence construction, punctuation, and conventions for graphics, it is very likely that you will need to look at one of these reference books to acquaint yourself with the terminology that we use to talk about these subjects. Don't think of it as time wasted; this familiarity will make it possible for you to use the brief rules and guidelines that follow, and in the end that will be the most efficient way for you to eliminate certain kinds of problems (which cause confusion and reduce your credibility) once and for all from your writing.

For convenience, the principles appear in three groups. You will find that each of the three groups has particular applicability to one of the three goals, but in practice the goals and the means of achieving them aren't entirely separable. The goals of precision and clarity both require correctness of grammar and syntax, and many techniques for achieving conciseness rely on your understanding of the rules that govern English sentence structure. Thus you will find principles under each heading that might well go under one of the others, or both.

Editing checklists based on these principles appear as **Appendix E.**

Thirteen Guidelines for Producing Clear Sentences

For grammatical terminology, see any standard grammar of English.

1. **Use the active voice in preference to passive voice whenever you can.**

   Don't Write   The **results have been obtained** through numerous tests conducted in our lab.

   Try   **We obtained the results** by conducting numerous tests in our lab.

   Don't Write   Corrosion of steel in a saline environment **is inhibited** by alloying it with chromium.

   Try   Alloying steel with chromium **inhibits** corrosion of the steel in a saline environment.

2. **Use straightforward subject-finite verb relationships in preference to incorporating miscellaneous material in modifying phrases.**

   Don't Write   **By alloying** steel with chromium, **corrosion** in a saline environment **can be inhibited.**

Try   **Alloying** steel with chromium **inhibits corrosion** of the steel in a saline environment.

Don't Write   **Being** a recyclable material, polyethylene **is suitable to be used for** this purpose.

Try   Polyethylene, because **it is** a recyclable material, **suits** the purpose.

3. **Avoid unsupported demonstrative and relative pronouns ('this,' 'that,' 'these,' 'those,' and 'which').**

Don't Write   Ceramic tile has high strength in compression, good insulating properties, and excellent resistance to wear. **This** makes is a good choice for the exterior surface.

Try   Ceramic tile has high strength in compression, good insulating properties, and excellent resistance to wear. **These three properties** make it . . .

Don't Write   Studies testing the efficiency of the engine under normal operating conditions show that the engine is less efficient and more costly to run than the designers had anticipated, **which is why** they have proposed modifications.

Try   Studies testing the efficiency of the engine under normal operating conditions show that the engine is less efficient and more costly to run than the designers had anticipated. **Consequently**, they have proposed modifications.

                 Studies testing the efficiency of the engine under normal operating conditions show that the engine is less efficient and more costly to run than the designers had anticipated. **Because of these flaws**, they have proposed modifications.

4. **Put modifying elements next to the things they modify, especially relative clauses and the qualifiers 'only,' 'all,' and 'not.'**

Don't Write	**A design** was produced by our team **that satisfied all the criteria**.
Try	Our team produced **a design that satisfied all the criteria**. **A design that satisfied all the criteria** was produced by our team (but see guideline 1 above).
Don't Write	The intraocular shunt is **only** helpful for patients who have mild glaucoma damage in one eye.
Try	The intraocular shunt helps **only patients** who . . .
	The intraocular shunt helps patients who have **only mild glaucoma damage** . . .
	The intraocular shunt helps patients who have mild glaucoma damage in **only one eye**.
	The intraocular shunt **can only <u>help</u>** patients who . . . ; it cannot reverse the damage or cure the disease.

5. **Use verb tenses accurately and avoid unnecessary use of modals.**

Don't Write	When the temperature and pressure **have reached** the designated levels, crosslinking **occurs**, and the material **will develop** strong bonds, which **would improve** the wear properties of the final product.
Try	When the temperature and pressure **reach** the designated levels, crosslinking **occurs**, and the material **develops** strong bonds. These bonds **improve** the wear properties of the final product.

6. Avoid existential constructions and extraposition.

Don't Write **There was a modification suggested by the client** that would increase the cost but improve several important features.

Try **The client suggested a modification** that would increase the cost but . . .

The client's suggested modification would increase the cost but . . .

Don't Write **It is likely that** the present electrode array could be replaced by conical electrodes with good results.

Try **Replacing** the present electrode array with conical electrodes **would probably produce** good **results.**

7. Avoid noun strings.

Don't Write We are presenting our **enhanced solid waste treatment facility plan** to the **Ann Arbor government waste reduction committee decision making body**.

Try We are presenting our plan for a facility that treats enhanced solid waste to the decision-making body of Ann Arbor's committee for reducing government waste.

We are presenting our enhanced plan for a solid-waste treatment facility to the decision-makers of the Ann Arbor City Government's committee for the reduction of waste.

We are presenting our plan for an enhanced facility for the treatment of solid waste to the members of the Ann Arbor Committee for Reducing Government Waste who are responsible for making decisions.

8. Avoid elliptical clauses that dangle or lack a logical agent.

Don't Write	**When implanting** the cochlear device, it must be carefully placed through the round window by a skilled surgeon.
Try	**During implantation** of the cochlear device, a skilled surgeon must place the device carefully through the round window.
Don't Write	**If properly implanted** within the inner ear, the surgeon can adjust the position of the electrodes to maximize contact with the intact hair cells.
Try	**If the device is properly implanted** within the inner ear, the surgeon can adjust . . .
	If the surgeon has properly implanted the device within the inner ear, he (or she) can adjust . . .

9. Make sure any comparative constructions are complete and logical.

Don't Write	The **control system** of the redesigned product **is** significantly **more complex compared to** the **prototype**, but, **like most redesigns, features** are incorporated that will make the final product **more marketable**.
Try	The **control system of the redesigned product is quite complex compared to that of the prototype**, but the **product, like most products that are redesigned,** incorporates features that will make **it more marketable than the earlier version** would have been.
	The redesigned product's **control system** is significantly **more complex than that of** the prototype, but **like most such products, the redesigned one** incorporates features that **will increase its marketability over that of** the prototype.

10. **Put important information into the sentence cores rather than into subordinated elements.**

> **Don't Write**   If there is a storm, assuming it is accompanied by lightning, it is possible that the sensitive equipment that we have installed will be damaged.
>
> **Try**   Lightning from a storm could damage the sensitive equipment that we have installed.

11. **Make sure that personal pronouns are easy to connect with one and only one referent (antecedent).**

> **Don't Write**   Though studies reporting the results of tests on the performance and energy efficiency of the engine under normal operating conditions show **it** to be worse than the designers had anticipated, **they** maintain that **it** is possible that **they** were erroneous and **it** is within acceptable limits.
>
> **Try**   Though studies reporting the results of tests on the performance and energy efficiency of the engine under normal operating conditions show that the engine's performance is worse than its designers had anticipated, the studies maintain that the results may be erroneous and performance may be within normal limits.
>
> Though studies reporting the results of tests on the performance and energy efficiency of the engine under normal operating conditions show that both performance and efficiency are worse than the designers had anticipated, the designers maintain that the studies may be erroneous and the engine may perform acceptably.

12. Use parallel constructions in lists and coordinated structures, and be careful not to telescope series.

Don't Write   The improved device has several advantages: it has lower bearing stresses than previous models, higher resistance to wear, and transmits forces in nearly the same way the natural structures do.

Try   The improved device has two advantages: one, it has lower bearing stresses than previous models and higher resistance to wear and, two, it transmits forces in nearly the way the natural structures do.

The improved device has several advantages: its bearing stresses are lower than those of previous models, its resistance to wear is higher than that of previous models, and its transmission of forces is very similar to that of the natural structures.

The improved device has several advantages. It has lower bearing stresses and higher resistance to wear than previous models, and it also transmits stresses much as the natural structures do.

Don't Write   The process is complex and involves the addition of a catalyst, reduction of the original mixture to half its volume, separation of the constituents through precipitation, release of carbon dioxide and limited amounts of nitrides.

Try   The process is complex; it involves (1) the addition of a catalyst, (2) the reduction of the original mixture to half its volume, (3) separation of the constituents through precipitation, **and** (4) release of carbon dioxide and limited amounts of nitrides.

13. **Make sure that your punctuation separates and groups sentence elements appropriately.**

Don't Write   The fuel cell, an environmentally-friendly alternative to internal combustion engines has not yet received the support of big corporations, which it will need if it is going to take its place alongside other energy-producing technologies most likely because of the lack of an infrastructure, that would allow drivers to rely on vehicles, powered by the cells.

Try   The fuel cell, an environmentally-friendly alternative to internal combustion engines, has not yet received the support of big corporations **that** it will need if it is going to take its place alongside other energy-producing technologies, most likely because of the lack of an infrastructure that would allow drivers to rely on vehicles powered by the cells.

The fuel cell, an environmentally-friendly alternative to internal combustion engines, has not yet received the support of big corporations— which it will need if it is going to take its place alongside other energy-producing technologies—most likely because of the lack of an infrastructure that would allow drivers to rely on vehicles powered by the cells.

Eight Guidelines for Producing Concise Prose

1. Reduce relative clauses to adjective phrases or appositives when you can.

   Don't Write   The design **that will be most successful** will incorporate all the features **that are essential**.

Try	The **most successful** design will incorporate all the **essential** features.
Don't write	The implant, **which is a bone plate that is made of stainless steel**, will remain in the body, **which is a factor that necessitates biocompatibility**.
Try	The implant, **a stainless steel bone plate**, will remain in the body, **a factor necessitating biocompatibility.**

2. Use single-word verbs rather than '*be* + adjective' phrases when you can (e.g., 'depends on' instead of 'is dependent on').

Don't write	The design's final form **is an outgrowth** of all the factors that we **took into consideration.**
Try	The design's final form **grows out of** all the factors that we **considered.**
Don't write	We **are suspicious** that our method **was different** from the one the experimenters had used because our results **were** quite **surprising** to us.
Try	We **suspect** that our method **differed** from the one the experimenters had used because our results **surprised** us greatly.

3. Use active voice rather than passive voice whenever you can.

Don't write	The results **were found** in a study that **was conducted** in 1998 by ISR.
Try	The results **appeared** in a 1998 ISR study.
	We **found** the results in a study that ISR **conducted** in 1998.

4. Use finite verbs rather than verbals whenever you can.

Don't write	By **calling** this number, it is possible **to obtain** information.
Try	**Call** this number for information.
	If you **call** this number, you **can get** information.
	Information is available at this number.

5. Avoid existential constructions and extraposition.

Don't write **There was** an experiment that demonstrated the hypothesis.

Try An experiment demonstrated the hypothesis.

Don't write **It is** essential that scientists be objective.

Try Scientists must be objective.

Objectivity is essential for scientists.

6. Use action verbs rather than forms of *to be* whenever you can, and be alert for opportunities to reduce and embed sentences of the form *subject–be-predicate nominative/adjective.*

Don't write Volta **was** a prominent **figure** in the early history of electrical engineering, and he performed some seminal experiments.

Try Volta, an important early figure in the history of electrical engineering, performed some seminal experiments.

Volta, who performed some seminal experiments, figured prominently in the early history of electrical engineering.

Don't write The results of the tests **were indicative** of some degradation of the material and they **are** recorded in the appendix.

Try The results of the tests, **recorded** in the appendix, **indicate** some degradation of the material.

The results of the tests, which **indicate** some degradation of the material, **appear** in the appendix.

7. Avoid thoughtless repetition of fixed phrases (e.g. 'viable alternative,' 'at this point in time,' and so on), redundancies ('in my opinion, I think') and empty words ('basically.)'

8. Avoid inflated diction (e.g. 'in terms of' **instead of** 'in,' 'of,' or 'about,' 'due to the fact that' or 'it being the case that' **instead of** 'because,' and so on).

Five Guidelines for Producing Good Diction

1. **Never use a word or phrase whose meaning you're not really sure of.**

   Example: *In lieu of the current budget constraints, we will need to make every effort to contain costs.

   The writer misunderstands *in lieu of* (French for 'in the place of') as 'in view of,' with nonsensical results. If you don't really know what a phrase in a foreign language means, don't use it.

   Example: *The budget constraints require us to modify the proposed design; as such, we will be submitting a revised proposal within the week.

   The writer misunderstands *as such* (a phrase meaning 'as (an example of) the previously-mentioned thing') as 'therefore,' 'as a result,' or 'so' and ends up sounding inflated and silly (*As (an example of) the proposed design, we . . . '?). *Such* is a pronoun, and it has to refer to a noun that precedes it, usually at the end of the previous clause. If you don't understand how an English phrase works, don't use it.

2. **Use the shortest and simplest word that will accurately convey your meaning; use polysyllabic words (apart from essential technical terms) only when you really need to.**

   Example: *The prototype of our robotic device exhibited a propensity toward retrogressive ambulatory motion.

   The writer seems to be mistaking length for informational content, but if the sentence means 'The prototype of our robot tended to walk backward,' isn't it more effective to say that? Note that 'prototype' is perfectly appropriate in the context.

3. **Never use an abstract term or generalizing pronoun when you can use a concrete term or a specific noun instead.**

   Example: *Many argue that some things about modern society are major factors in the current situation with stem cell research.

The writer's meaning is obscured because most of the content-bearing words in the sentence are vague. You need to decide whom or what you're actually thinking about when you write 'many' and what you actually mean when you write 'things' or 'situation.' All of the sentences below could replace this sentence, but as you can see, they have very different meanings. Concrete nouns and specific verbs put meaning into sentences—use them.

Many sociologists argue that the character of modern society influences societal attitudes toward stem cell research.

Proponents of stem cell research argue that societal ambivalence about what constitutes 'human life' affects voters' support for this research.

Opponents of stem cell research claim that modern society's casual acceptance of abortion contributes to the current trend of support for this research.

4. **Never add a word, phrase, or clause unless you know what it's doing in the sentence. Check coordinated elements to make sure that any modifiers you add work with both/all of them. Test coordinated objects to make sure that both/all of them work with the verb you've used. Test qualifiers to see whether they're actually qualifying the element you meant to qualify.**

5. **Avoid writing anything that sounds unnatural when you say it aloud.**

Ten Rules for Punctuating to Produce Clarity

1. Use commas at **both ends** of movable units (dependent clauses, parentheticals, nonrestrictive relative clauses, adjective phrases, and appositives, conjunctive adverbs, and some qualifiers) when they occur **within** your sentences. If these sorts of elements occur at the beginning or end of your sentence, you need only one comma, because the end punctuation marks the other end.

2. Use commas to separate items in lists of more than two items, but don't separate coordinated structures consisting of only two items.

3. Use a comma at the end of an independent clause that you're joining to another independent clause with a coordinating conjunction (*and, or/no, but,* and *so*). Don't put a comma after a coordinating conjunction, and don't put a comma after a subordinating conjunction.

4. Don't separate your subject from your verb or your verb from its object with any mark of punctuation (unless you're inserting a moveable element with commas at both ends).

5. Use a semicolon at the end of an independent clause that you're joining to another independent clause without a coordinating conjunction. Don't confuse conjunctive adverbs, which are moveable elements, with coordinating conjunctions.

6. Use semicolons to separate the elements of a list if and only if the elements are complex phrases that have commas in them.

7. Use a colon at the end of an independent clause that introduces a list or amplification. Don't use colons between core sentence elements.

8. Use italics or quotation marks to indicate a word or phrase used as a term rather than in its actual meaning (i.e., *The word 'receive' is often misspelled.)* Use quotation marks to indicate that you are using a word ironically (these are sometimes called 'sneer quotes.')

9. Use hyphens to show that two words are functioning as a single grammatical unit (like two-part adjective phrases) rather than as separate items.

10. When you insert quoted material into a sentence, make sure that the resulting sentence follows the regular rules for punctuation. Put end punctuation inside the quotation marks if the quotation ends the sentence (but it probably won't, because you'll

need your parenthetical citation or bracketed reference). Put
your end punctuation after your closing parenthesis or bracket).

The following rules aren't so much for clarity as to save you from
embarrassing errors.

11. Use an apostrophe when you need to add an *–s* to indicate pos-
session, but not when the *–s* is organic (i.e., *the sun's temperature* but **not** *it's temperature*; *the book is my sister's* but **not** *the
book is her's*). Use an apostrophe when you omit a letter or letters in contractions. Don't use it when you pluralize words, numbers, or letters in use as symbols or acronyms.

12. Capitalize proper names consistently within a text, and don't
capitalize words that aren't proper names unless you have a
good reason to do so.

C

STRATEGIES FOR
DEVELOPMENT
OF EFFECTIVE GRAPHICS

You will very often need to incorporate graphics, which include pie and flow charts, line and bar graphs, tables, and illustrations, into technical reports and presentations. Your purpose for including a graphic within your report can vary. The reason may be to emphasize a point you are making within the accompanying text, to offer quantitative support for a claim, to depict a relationship among parts of a complex object or system, or to help the audience visualize that object or system. You can develop effective graphics by applying a few simple principles related to the content and to the design.

When developing graphics for technical reports and presentations, you must both accurately *and* clearly represent the data or information for your intended audience. If you are careful to do so, then the audience will more easily be able to understand your reason for showing the information. A set of simple guidelines related to both content and design is included in this appendix.

Guidelines For Effective Graphics

1. **Represent the data in your visual accurately (use correct scales and ranges of data)**

Presenting data in a graphic that falsely or misleadingly represents the nature of that data harms the credibility of the report or presentation in which it occurs (see **Appendix A** for a discussion of the importance of credibility). Furthermore, if you have gathered data from the results of other people's research, you will need to cite these sources. Failing to cite your sources would falsely suggest that you had generated the data yourself.

Pay particular attention to the range of data and the scales you use to present the information. You should be aware of possible misinterpretation of information that occurs when the ranges and scales you select vary. For example, the line graph (Figure 1A) on page 137 shows what appears to be a trend of increasing profit. From the audience's perspective, the graph may suggest further assumptions about the information (for example, that more current data had not yet been collected).

However, if the graph were to present the entire range of data that had been collected (Figure 1B), the audience would see that the profitability value has actually flattened out in recent months.

2. **Use the style of visual best suited to represent the data**
    - **Pie charts [Chart 1]** show fractions or percentages of a whole number.
    - **Line graphs [Figures 1A, 1B, and 2.1]** use data points to show trends over a period of time.
    - **Bar graphs [Figure 2.2]** can sometimes show trends, but are best used to highlight specific data points and permit comparison.
    - **Illustrations [Figure 2.3]** show objects or depict systems and physical relationships

Note that the degree of precision you need in order for your graphic to support the points you are making in your text will vary, and this consideration will influence your decision as to the most appropriate graphic. For example, bar charts can quickly become cluttered and ineffective if many numbers are superimposed on the bars, so if you need precise numbers, you may want to choose another graphic to display your data.

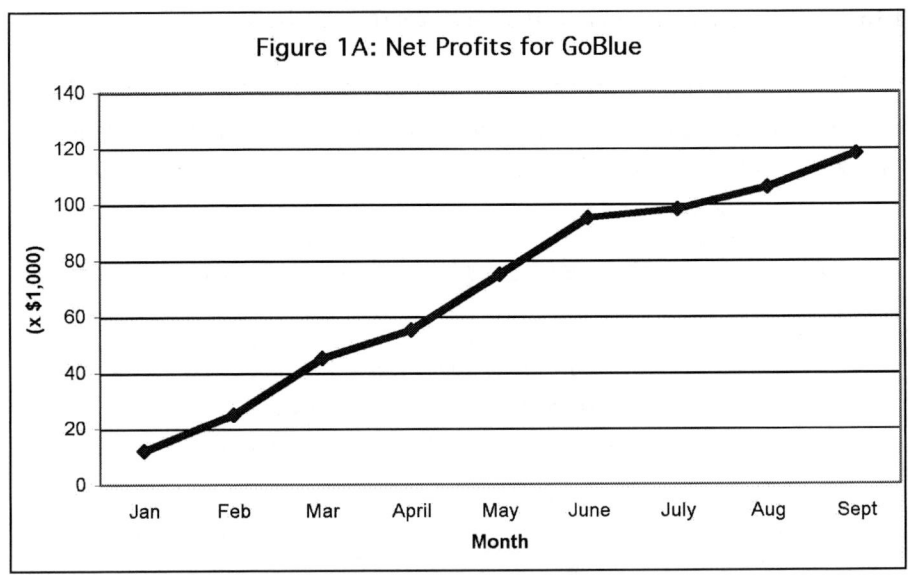

Figure 1A: Net Profits for Go Blue, Inc. for 2007

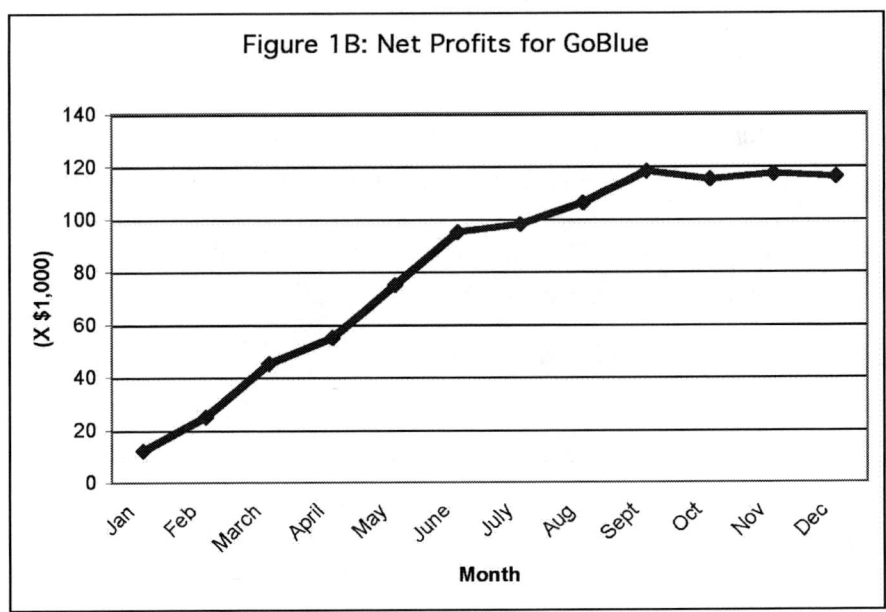

Figure 1B: Net Profits for GoBlue, Inc., for 2007

Similarly, pie charts do not easily show small differences between numbers or quantities, and if the reader must know exact percentages to get the point you want to make, you will have to supply them, possibly in an accompanying table.

Examples of several types of graphics are given below:

Chart 1: Breakdown of college costs($) illustrating that tuition accounts for more than 50% of the total expenses

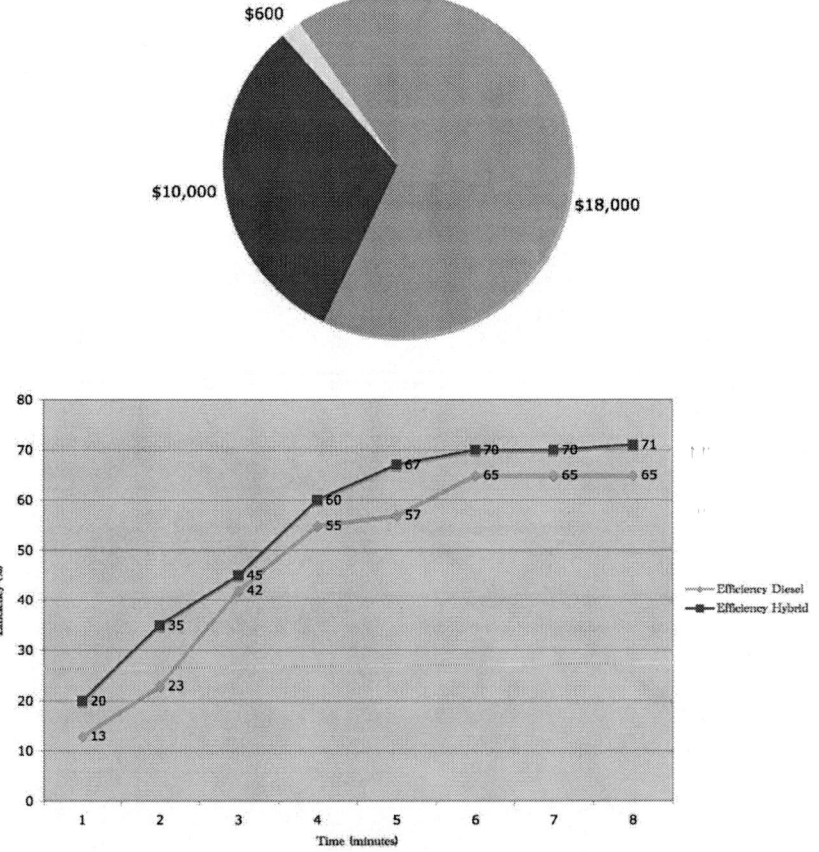

Figure 2.1: Efficiency (%) of a diesel engine compared to the efficiency of a Hybrid engine

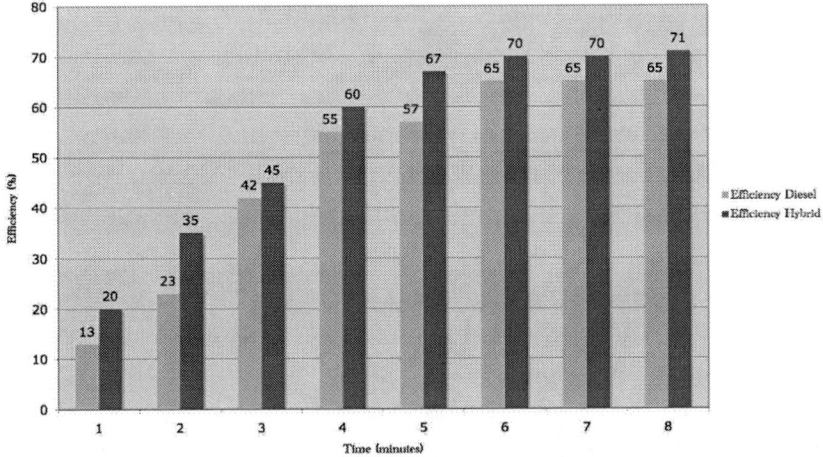

Figure 2.2: Efficiency (%) of a diesel engine compared to the efficiency of a Hybrid engine

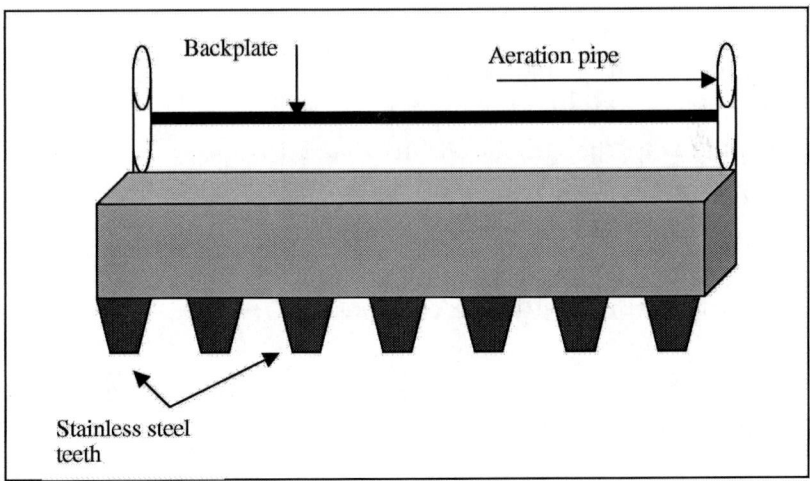

Figure 2.3: Concrete pier anchor design with 7 stainless steel teeth and two aeration pipes

When it comes to pictures, consider whether a drawing may be better for your purpose than a photograph. A photograph can be a more accurate depiction, but a line drawing can be made to focus on only those aspects of an object that you must describe. So, while a photograph may be more striking and detailed, it may not serve you as well in communicating a particular point or showing a particular relationship.

3. Refer to and explain the graphic in your text and place it physically near the textual reference within a written report to improve its usability.

If you fail to direct readers to your graphic, they may not use it appropriately (or at all); if the graphic is placed far away from the related text, readers may encounter it too late for it to help them understand the point you were trying to make.

To illustrate, if you were to include the figure represented by the text box at the right, you would begin your explanation of it here with a statement like "As Figure 1 shows . . . " or "Figure 1 illustrates . . . " Introduce the graphic as soon as it would benefit the reader to look at it. Some people recommend that for a professional appearance, you should wrap the graphic with your text; in any case, the placement of the text and graphic should be rational.

4. Use descriptive titles or captions.

The title or caption should clarify what the figure is showing. For example, a simple title such as

Figure 4: Product Assessment

is not as descriptive, and therefore not as meaningful to the reader, as at title such as

Figure 4: Product Assessment for an I-book G1 laptop computer evaluating the recycling and disposal life stages

The detail, however, should go into the text. Captions should be no longer than they need to be to clarify the content of the graphic (in some cases, they may run to a sentence or two).

5. Use labels, as necessary.

For example, in Figure 2.2 (the bar chart), which shows a comparison of engine efficiency, each individual data value is shown above each of the bars. This type of detail in labeling may be helpful to the reader.

6. Highlight your main points (use bold font type, color or arrows)

Use graphic indicators to show the hierarchy of the information.

Lab Equipment Cost	Purchase 2006	Purchase 2009
Stress-Strain Materials Test Machine	$32,000	$42,000
Scanning Electron Microscope	$150,000	$200,000
Total Cost	**$182,000**	**$242,000**
Difference in Cost		**$60,000**

Table 6.1: Lab equipment cost comparison showing that the total cost of the equipment would be greater by $60,000 in 2009

7. Maintain a uniform layout

Develop and maintain a uniform format for the visuals by using similar font types and sizes, similar sizes of graphic, and similar labeling types and styles within a report or presentation; uniformity increases the cohesion and professional appearance of your report.

8. Number your graphics (Examples: Figure 1.2, Table 3.2)

You must number your graphics in order to refer to them clearly within your text.

9. **Show references on the graphic**, if the data or graphic is copied from a source. Provide the necessary citations on your graphic. If you copy it, cite the source of the graphic; if you construct it yourself, cite the sources of the information you include in it.

10. **Avoid visual clutter:** this term refers to excessive information or unnecessary effects that do not help to clarify the main message of your graphic. You should avoid putting vast quantities of information into a graphic so that the audience will more easily be able to focus on the main point(s) that you are making. The figure below shows a 3-D bar graph that is not easy for the reader to use. The data points are numerous and difficult to read, and the trends are difficult to identify because of the 3-D effect. This is an example of a graphic made relatively unhelpful to a viewer by the inclusion of excessive information.

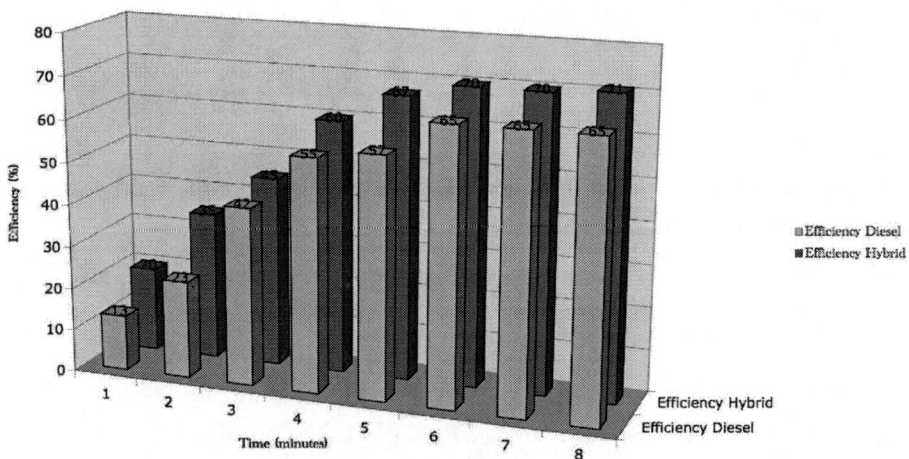

D

GUIDELINES FOR
PROFESSIONAL USE
OF E-MAIL

Many people seem to think that e-mail messages, because we often exchange them with our friends, are invariably informal in style and need not follow the rules governing other written communications. Actually, the level of formality depends on the relationship between sender and receiver, not on the medium itself. Consider these two things:

- in a professional setting, e-mail may be the main communication channel you have with managers or leaders in the company;
- e-mail messages are often printed out and filed as part of the documentary history of a particular project or approach to a problem.

So although we do use e-mail casually to communicate with our friends, when we use it in professional contexts, what we write has to observe both rules for writing and rules of etiquette. Some of the latter are still in flux, but a survey of many resources, including internal guidelines put out by major industries, confirms that the ones listed below are now fairly well established.

1. Use an appropriate salutation and close.

    Address the recipient by name, but don't assume that you can use people's first names; do so only when you know it won't offend or irritate your reader. 'Dear [name]' is perfectly proper; you may also use 'Hello [name]' or, when you're very sure that the communication is fairly casual, '[Hi [name].' Before the name, use an appropriate title: *Dr., Mr., Ms.*, and *Prof.* are likely choices. In some instances, you may use 'Dear [First name] [Last name]'– this might be necessary if you can't tell the recipient's sex from the first name or if you've had some friendly contact with the recipient but don't feel comfortable using his or her first name. Use a formal close, like 'Sincerely,' if the overall communication is formal; if it's slightly more casual, you can use a phrase like 'Best regards.' Under the close, give your full name unless the recipient would recognize your name without any additional context (and always if your login ID is uncommunicative.) Some people recommend providing your affiliation as well.

2. Keep your messages short, and keep units within messages short.

    Break relatively longer messages into short blocks; they're visually less daunting than long, dense paragraphs. Evidence suggests that very messages are often simply not read. **If your message must be long, provided a summary at the beginning.**

3. Retain the thread or use informative subject lines.

    Some people disagree, saying that preserving previous messages adds unnecessary length, but a majority of the sources we surveyed agreed that busy people need the context thus provided to read the message with full understanding.

4. Use Standard English grammar, punctuation, and spelling.

    Every list of rules we've seen mentions this. Many people think that e-mail doesn't have to follow the same conventions as paper documents. Not so! Engineering professionals and business leaders alike say that poor spelling, irregular use of capital letters, incomplete sentences, and absence of punctuation in e-mails they receive produce at best irritation and at worst confusion. Over and over we see 'computers have spelling and grammar checks; why don't people use them?' in the comments of executives asked about what bothers them about e-mails their employees send them. **PROOFREAD YOUR MESSAGES BEFORE YOU SEND THEM –AVOID THE DREADED 'ONO-SECOND.'**

5. Avoid using abbreviations that may not be standard, and use graphic devices carefully.

   Most people will recognize 'BTW' or 'FYI,' but the vast abbreviation lexicon of the IM realm is unfamiliar to many. In general, avoid using all caps; people perceive it as overbearing. Some people maintain that emoticons are entirely unsuitable for business and professional communication; others argue that the sparing use of them is permissible in the most informal business messages.

6. Get right to the point, but watch your tone.

   Studies show that many e-mail messages are perceived as brusque or even rude, even when the writer's intent was merely to be concise. Use *modals, conditionals,* and *questions* to soften the tone and avoid sounding peremptory or abrupt, and use *imperatives* very sparingly. E.g., instead of 'Please review the attached material and get back to me ASAP,' write 'Could you review . . . If you could get back to me ASAP. I'd appreciate it [etc.].' Don't overdo the buffers, but don't lead off with your request, usually.

7. If you include attachments, specifically say so in the text of your message.

   Tell the reader both the name of the file and the format it's in. Avoid generic names for your attachments (like 'My Proposal' or 'Resume').

8. Avoid successive forwarding and long copy lists.

   Studies have shown that readers are often irritated by the quantity of extraneous material that precedes the message in such cases.

9. Don't flood people's inboxes with messages.

   Think about what you're writing, so that you won't have to follow every message with another one containing something you forgot.

E

CHECKLISTS FOR ASSESSING, EDITING, AND REVISING YOUR DOCUMENTS

Editing Checklist 1

- ❑ Document follows the prescribed format
- ❑ Document clearly states its scope and purpose at or near the beginning
- ❑ Document provides an easy-to-follow overview
- ❑ Important terms are explained early and terminology is consistent from section to section
- ❑ Document has clear section delineations (headings, spacing, indentation)
- ❑ Figures are consistently numbered and clearly labeled
- ❑ Graphics are appropriately captioned or titled
- ❑ Figures, tables, and graphs are referred to and explained in the text
- ❑ Sources are consistently and accurately documented
- ❑ References are presented in prescribed style
- ❑ Any necessary appendices are included and clearly titled

Editing Checklist 2
- ❑ Most sentences begin with substantive elements
- ❑ Concrete language is used as much as possible
- ❑ Punctuation clearly marks movable phrases
- ❑ Punctuation clearly indicates independent clauses
- ❑ Slang, dialect, and non-standard forms are absent
- ❑ Active voice is used as much as possible
- ❑ Modifying elements are next to the things they describe
- ❑ Precise numbers or concrete nouns replace generalizing pronouns as much as possible

Editing Checklist 3
- ❑ Sections and paragraphs are usually organized from general to specific
- ❑ Sentences in paragraphs are related to topic sentence
- ❑ Logical connectives show the relationships between ideas
- ❑ Paragraphs are focused and unified
- ❑ Most paragraphs begin with clear topic sentences
- ❑ Reasonable, defensible claims are clearly presented
- ❑ Adequate and appropriate supporting material is provided for claims
- ❑ Logical fallacies are avoided
- ❑ Material taken from secondary sources is clearly distinguished from your own words

F

FEATURES OF FORMAL REPORTS

A formal report differs from an informal report or memorandum in two main ways: 1) instead of a memo heading, it has a title page that gives the same information and often some additional information, and 2) it has a table of contents that precedes the Foreword and Summary section. If you use an Executive Summary, the table of contents will follow it. It may also have some other elements (these typically include an abstract and separate lists of illustrations and tables; sometimes it may include acknowledgements or a preface). It is often presented like a book, with binding and new pages for each new section, especially if it's going outside your organization.

A title page typically contains the following information, and we recommend this order:

The title of the report

The names of the writers (in alphabetical order)

The date the report is submitted

The names, titles and positions of the recipients

Particular projects may call for additional information on a coversheet. You will see examples of coversheets following this pattern as part of the proposal and design report in this book.

A table of contents should follow these rules: the items on the left-hand margin must correspond exactly to your headings and subheadings; the page numbers must be neatly aligned; the headings must be easy to line up visually with the page numbers, and the entire thing must be spaced so as to make the reader's job of finding the right page number for a particular section very easy. For models, see the examples included as part of the proposal, design report and research report in this book.

The fact that you are now going to have a table of contents has some implications; you have to make some decisions about how you're going to divide and label the contents for maximum reader-friendliness. You may choose to number your sections and use indentation to signal levels.

ACKNOWLEDGMENTS

The authors wish to thank the following colleagues and friends for their helpful suggestions and criticisms: Leslie Olsen, J.C. Mathes, Dwight Stevenson, Mary Gilbert, Ken Alfano, Robin Roots, Walburga Zahn, Tom Bowden, Peter Nagourney, Rob Sulewski, Rod Johnson and Elaine Wisniewski.

We further wish to thank several of our students, past and present, who allowed us to adapt their written work for use in this book: Colleen Budd, Shari Hannapel, Nicole Klauza, Michael Krug, Timothy Roberts, and Stepan Tikhonov.

We're grateful to Professors Gary Herrin and James Holloway, past and present Associate Deans for Undergraduate Education, University of Michigan College of Engineering, and Professor Toby Teorey, Director of Academic Programs, Undergraduate Education, University of Michigan College of Engineering, for their support of our work.

We also wish to acknowledge our debt to the following resources, from which we have gathered ideas, advice, and general principles:

Finkelstein, Leo (2005). *Pocket Book of Technical Writing*. 2nd edition. McGraw-Hill, Inc.

Flesh, Rudy (1948). The Art of Readable Writing. New York: Pocket Books.

Frees, Edward W., and Robert B. Miller, Designing Effective Graphs. Retrieved from the World Wide Web, January 2, 2007, at www.actuaries.asn.au

Mathes, J.C., and Dwight W. Stevenson (1991). *Designing Technical Reports: Writing for Audiences in Organizations.* 2nd edition. Macmillan Publishing Company.

Olsen, Leslie A., and Thomas N. Huckin (1991). *Technical Writing and Professional Communication.* 2nd edition. McGraw-Hill, Inc.

Swales, John M., and Christine B. Feak (2001). *Academic Writing for Graduate Students.* Ann Arbor: University of Michigan Press.